An Introduction to Safety Grounding

An Introduction to Safety Grounding

Asser A. Zaky

CRC Press is an imprint of the
Taylor & Francis Group, an **informa** business

First edition published 2022
by CRC Press
6000 Broken Sound Parkway NW, Suite 300, Boca Raton, FL 33487-2742

and by CRC Press
2 Park Square, Milton Park, Abingdon, Oxon, OX14 4RN

Library of Congress Cataloging-in-Publication Data
Names: Zaky, Asser A. (Asser Aly), 1932- author.
Title: An introduction to safety grounding / Asser A. Zaky.
Description: First edition. | Boca Raton, FL: CRC Press, 2021. | Includes bibliographical references and index. | Summary: "Protective or safety grounding is of vital importance for the protection of individuals from electric shock. The objective of this book is to give the reader a better understanding of safety grounding, why it is needed, where it is needed, and what are the requirements which must be met in order to have an effective grounding system"—Provided by publisher.
Identifiers: LCCN 2021004135 (print) | LCCN 2021004136 (ebook) |
ISBN 9780367758714 (hbk) | ISBN 9780367759278 (pbk) |
ISBN 9781003164630 (ebk)
Subjects: LCSH: Electric currents—Grounding. |
Electric wiring—Insurance requirements.
Classification: LCC TK3227 .Z35 2021 (print) | LCC TK3227 (ebook) | DDC 621.31/7—dc23
LC record available at https://lccn.loc.gov/2021004135
LC ebook record available at https://lccn.loc.gov/2021004136

ISBN: 978-0-367-75871-4 (hbk)
ISBN: 978-0-367-75927-8 (pbk)
ISBN: 978-1-003-16463-0 (ebk)

Typeset in Times
by codeMantra

Contents

Preface

It has often been said that grounding is part science and part art. To a large extent this is true if we consider the art part to include the extensive experience accumulated over centuries, going back to Benjamin Franklin's experiments with kites and lightning and his invention of the lightning rod, which gave birth to the concept of grounding as a safety measure.

As with many engineering inventions, theory followed practice and the combined outcome of both theory and practical experience have been embodied in a number of national and international standards and codes of practice on safety grounding, which are revised from time to time to take into account the development of new materials, improvements in engineering practice, and new research findings.

Ever since grounding was used as a means of protection against lightning, its importance as a safety measure in electrical engineering applications, from generation to utilization, has been universally recognized and is today an obligatory safety requirement for all electrical installations and all nonbattery operated electrical equipment. Unfortunately the implementation of this requirement is more often than not overlooked in many developing countries. The reason for this is a mixture of ignorance and a laxity in the enforcement of safety measures. The majority of engineering curricula does not explicitly address safety grounding but emphasizes system grounding. It is not surprising therefore that engineers do not always distinguish between these two types of grounding. This lack of awareness applies also to technicians and wiring electricians. There certainly is a need for a wider understanding of safety grounding and its importance as a life-saving measure.

During years of teaching electrical power and high voltage courses I have always made it a point to include safety grounding as an integral part of the course. The present text is the outcome of these lecture notes enlarged and updated as far as possible to conform to present-day standards and practice.

The choice of material for a balanced textbook on grounding is not easy. A comprehensive treatment of each individual topic requires a book by itself; for instance, Tagg's classic book *Earth Resistances* is almost exclusively devoted to earth electrodes, Golde's *Lightning Protection* to the protection against lightning, Loeb's *Static Electrification* to the triboelectric effect to quote but a few. In the present text the topics and depth of treatment have been chosen with the prime objective of giving the reader a simple theoretical background to each topic and a "feel" for the design of practical grounding systems which is intimately and inexorably bound to existing standards, both national and international. Hence reference to such standards and their requirements is frequently made throughout the book.

The book addresses students and professionals interested to learn more about safety grounding other than that it is a piece of wire with which equipment is connected to ground, especially since the very numerous standards on every aspect of the subject are not readily accessible.

The book has eight chapters:

Chapter 1 gives the physiological effects which the magnitude of current and its duration has on the human body.

Chapter 2 deals with the resistance of an electrode to ground and the resistivity of the soil. Methods of measuring this resistance as well as the resistivity of the soil are given. The resistance area of a ground electrode is defined.

Chapter 3 deals with the different types of ground electrodes and the effect of their geometry and numbers on the resistance to ground.

Chapter 4 covers the concepts of step, touch, and transfer voltages and their safe values.

Chapter 5 is the core of the book and presents in some detail the components of a ground system, methods of improving soil resistivity, the types of welds and joints,the criteria for determining conductor cross-sections, galvanic corrosion, and a survey of the different grounding practices used at substations and the different types of grounding systems used for the protection of consumers.

Chapter 6 gives a concise treatment of the use and design of substation ground mats such that step and touch voltages remain within the safe limits and end with examples using a computer program for the design of ground mats.

Chapter 7 deals in some detail with static electrification and the types of electrostatic discharges (ESD) for which they are responsible, especially since such discharges have today acquired considerable importance in the electronics manufacturing industries. The chapter includes the various methods used to minimize or prevent the occurrence of such discharges.

Chapter 8 introduces the reader to the subject of lightning and lightning protection. Lightning of course is as old as the earth itself and to this day many aspects remain unsolved with research still continuing. However, the protective measures used so far have proved to be quite effective and are continuously being improved upon as evidenced by the recent changes brought about in almost all relevant standards.

It should be pointed out that the use of “ready-made” computer programs for the design of a ground system may be labor saving but it is the author’s opinion that a sound knowledge of the premises on which such programs are designed is of fundamental importance for their proper use.

It is hoped that the coverage provided in the present monograph will help students and engineers alike to better understand protective grounding and the standards involved and appreciate its importance and so champion its implementation in their field of work and thereby be blessed for a life-saving act.

In preparing this text recourse has been had to many standards and references, and wherever necessary, tables and figures have been reproduced with due acknowledgment of their source. In particular, the author would like to thank W.J.Furse & Co for permission to use a number of their figures and the NFPA for permission to use their simple lightning risk assessment procedure. The author thanks the International Electrotechnical Commission (IEC) for permission to reproduce information from its International Standards.[1] Thanks are also due to the publishing staff of T&F for their assistance and guidance throughout the preparation of this book.

1 Effect of Current on the Human Body

1.1 INTRODUCTION

There are essentially two types of grounding:[1]

1. System grounding
2. Protective or equipment grounding

In the first type of grounding the star points of the equipment may be solidly grounded or grounded through a resistance or an inductance (Petersen coil) according to the operating requirements of the network. These requirements depend on several factors such as the maximum permissible stress on the insulation, the magnitude of the short circuit current, and the overall protective characteristics of the network.

The second type of grounding, which is the subject of this book, has two objectives:

(i) To protect people (and animals) in the event of the occurrence of a fault to ground.
(ii) To protect buildings and installations against fire and lightning.

In order to protect people (operators, maintenance and repair technicians, and the public at large) against electric shock if they come in contact with metal parts which normally are not live and do not carry any current, it must be ensured that under fault conditions the potential of such parts does not rise to a value which would be considered dangerous to persons or give rise to leakage currents which, even if very small, can with time raise the temperature of the material through which it flows to a value sufficient to initiate a fire if the material is readily inflammable. Protection is provided by deliberate grounding of all metal structures, motor, generator and transformer frames, metal enclosures of all tools and control equipment, connection boxes, cable trays, and all other metal bodies which contain or are adjacent to electric circuits and which are within reach of any person.

Figure 1.1 shows the equivalent circuit of an electrical equipment connected to a supply source of voltage V. The resistances shown are as follows:

R_1: resistance of the insulation between live parts and equipment case,
R_g: resistance between casing and ground, and
R_b: resistance of a person's body to ground.

[1] The terms *grounding* and *earthing* are used interchangeably.

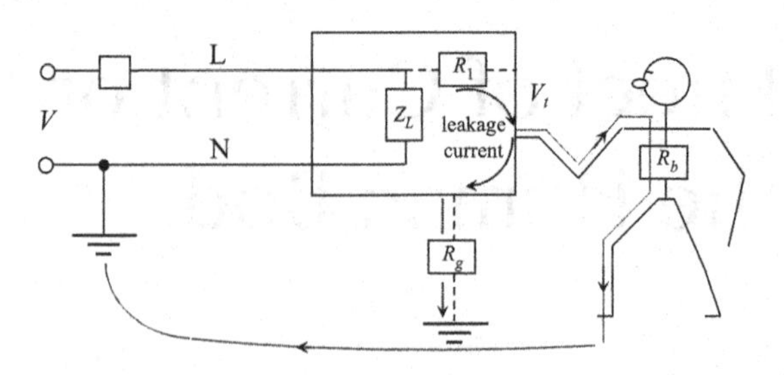

FIGURE 1.1 Isolated equipment.

The voltage V_t which appears on the case is

$$V_t = \frac{V R_g}{R_1 + R_g} \tag{1.1}$$

When a short circuit occurs between the live conductor and the case ($R_1 = 0$), V_t becomes equal to V and the current which flows through the body of a person who touches the frame is

$$I_b = V / R_b.$$

If the case were connected to a perfect ground such that $R_g = 0$ then the case would always be at zero potential. However, in practice the resistance to ground is never zero so that under fault conditions the case voltage is V and remains at that value until the protective device (fuse or circuit breaker) disconnects the supply (Figure 1.2). To make certain that the person who touches the case is not at risk the circuit must be disconnected within a specified time (see Section 1.3); since the operating time of the protective devices depends on the magnitude of the short circuit current, it is necessary to ensure that the resistance between the case and ground is sufficiently small to allow the passage to ground of a current whose value is sufficient to operate the

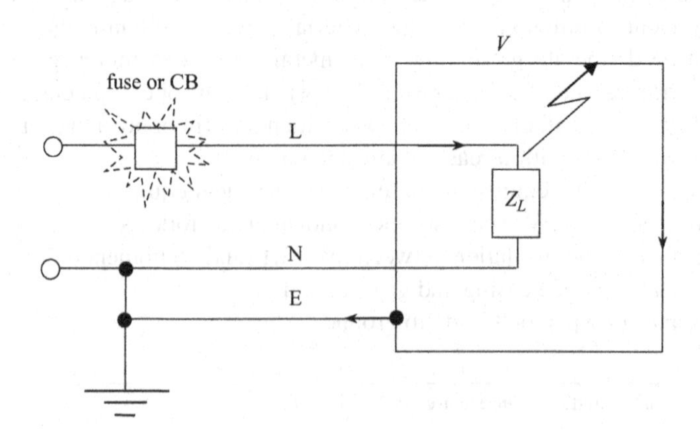

FIGURE 1.2 Earthed equipment.

protective device within the permitted time. To fulfill this requirement the resistance to ground must not exceed a certain value usually specified by national or international standards. This value varies between 1 and 25 Ω depending on the magnitude of the short circuit current.

As for the protection of buildings and installations against lightning strikes, this is accomplished by earthing systems especially designed for this purpose and its implementation is determined by how critical such a protection is. Protection of structures against lightning is dealt with in Chapter 8.

In all of the national and international specifications protective grounding of all equipment is an obligatory safety requirement irrespective of other considerations such as electromagnetic compatibility, for example. It is therefore of primary importance that there be a close cooperation between the engineering consultant responsible for the design of supply and grounding system of any installation and the designer of the electronic equipment to be installed in the building in order to choose the optimum system which will meet both the safety requirements and the compatibility requirements within the electromagnetic environment at the premises.

1.2 EFFECT OF ELECTRIC SHOCK ON HUMAN BEINGS

A person is subjected to an electric shock if he touches any live conductor or other metallic body while at the same time touching another grounded body or standing on a moist ground or is barefooted. Although the common belief is that the voltage is the cause of the shock, the consequences and severity of the shock depend on many factors, the most important of which is the magnitude of the current which flows through the body, the path of this current, and its duration.

Numerous studies and observations have shown that the effect of a low-frequency current (0–300 Hz) on the human body varies with the magnitude of the current which flows through the chest area. Table 1.1 shows a summary of the effects produced by both alternating and direct current magnitudes for durations of a few seconds. Note that the current magnitude is in milliamperes.

The minimum current which can be felt by human beings is one milliampere. Between 6 and 10 mA muscle control is lost so that a person cannot let go of any

TABLE 1.1
Effects of Current Magnitude on the Human Body

AC mA, 50 Hz	DC mA	Effect
0.5–1	1–4	Threshold of feeling
1–10	4–15	Pain
10–30	15–80	Let-go threshold
30–50	80–160	Muscular paralysis
50–75	160–300	Difficulty in breathing
75–250	300–500	(Ventricular fibrillation) death
> 250	> 500	Cardiac arrest and serious burns

electrified body held in his hand. As the current magnitude increases breathing becomes difficult and the muscles become paralyzed. These effects are not permanent and disappear if the current is switched off within a few seconds. Even if breathing stops the injured person can be saved from suffocation by artificial respiration. However, if the current is between 75 and 250 mA the electric shock is fatal. The reason for this is that within this current range the heart goes into a state know as ventricular fibrillation in which the heart muscles no longer contract in synchronism. This state is called cardiac arrest and it is more dangerous as it can only be reversed by specialized equipment only available in hospitals. Currents above 250 mA lead to cardiac arrest, cessation of breathing, and severe burns; however if the injured person is given immediate treatment resuscitation is possible.

In the case of direct current, the current magnitudes producing the above effects vary between two and four times the AC values as shown in Table 1.1.

At high frequencies the magnitude of the current required to produce the above effects increases with increasing frequency due to the skin effect. For example the threshold of feeling at 70 kHz is 100 mA and for frequencies higher than 100–200 kHz the effect is limited to a sensation of heat or to the occurrence of superficial burns.

1.3 EFFECT OF CURRENT DURATION

In order to prevent an electric shock from causing death the magnitude of the current must be less than that which causes ventricular fibrillation. As mentioned above the magnitude of this current varies between 75 and 250 mA (the actual value depends on the size of the body) if the duration of the current is a few seconds. Experiments and statistics indicate that the shorter the duration of the current the higher the current needed to cause fibrillation. Although there is complete agreement between investigators on the importance of current duration, there is no law or equation agreed upon internationally for relating the magnitude of the current causing fibrillation to the duration of that current. The most common relationship used is that arrived at by Dalziel[2] and modified more recently by others[3] on the basis of numerous experiments. This relationship is

$$I_b = k / \sqrt{t} \quad \text{ampere} \tag{1.2}$$

where t is the duration of the shock current in seconds and k is a constant whose value depends on the weight of the person:

$k=0.116$ for 99.5% of persons weighing 50 kg
$k=0.157$ for 99.5% of persons weighing 70 kg
If we assume that $k=0.116$ then,
$I_b= 116\,\text{mA}$ $t=1$ s
$I_b= 367\,\text{mA}$ $t=0.1$ s

[2] G.F. Dalziel, Dangerous electric currents, *AIEE Trans.*Vol.65, pp 579–585, and 1123–1124, 1946.
[3] J.G. Sverak, W.K. Dick, T.H. Dodds and R.H. Heppe, Safe substation grounding – Part I, *IEEE Trans. PAS*, pp 4281–4284, 1981.

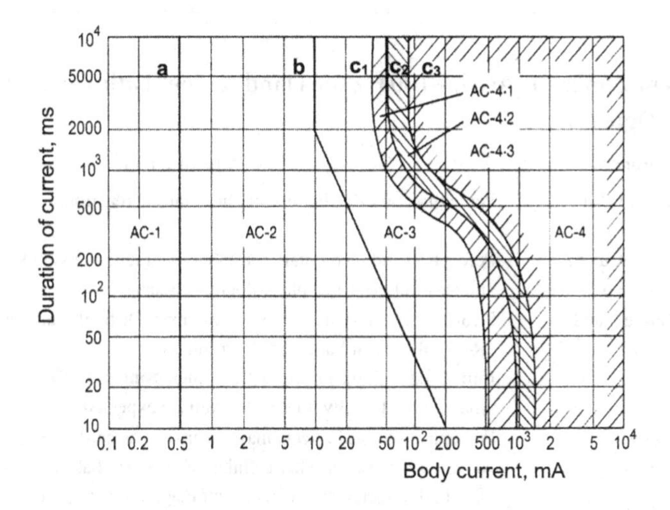

FIGURE 1.3 Conventional time/current zones of effects of AC currents (15–100 Hz) on persons for a current path corresponding to left hand to feet according to IEC 60479-1 (see Table 1.2).

Equation (1.2) is based on experiments in which the duration of the current varied between 0.03 and 3 seconds and on the assumption that the resistance of the human body is 1,000 Ω (see Section 1.4). This equation shows that higher shock currents are allowed if high-speed protective devices are used to disconnect the current and limit its duration. If the probability exists of prolonged shock duration without immediate aid, then to prevent the injured person from choking due to cessation of breathing, it is preferable to limit the maximum shock current to 25 mA. However, most designers prefer to limit this current to the let-go value (9 mA for men and 6 mA for women).

Figure 1.3 gives the physiological effects of alternating current (15–100 Hz) duration on the human body according to the International Electrotechnical Commission (IEC) standard,[4] and a summary of the time zones is given in Table 1.2.

1.4 THE ELECTRIC RESISTANCE OF THE HUMAN BODY

It is difficult to specify an exact value for the resistance of the human body since its value is determined by several factors such as age, sex, weight, general physical condition, extent of skin dryness, and the position of the body at the time of the accident. The total resistance of the body consists of two parts: the skin resistance and the internal resistance.

The skin resistance varies between 100 Ω/cm^{2}for wet skin and 3 × 10^{5}Ω/cm^2 for dry skin with values higher than that for people with coarse hands such as manual workers.

The internal resistance of the body is in the range 400–600 Ω between extremities—hand to hand, hand to foot, or foot to foot.

[4] IEC 60479-1 ed.1.0 "Copyright © 2018 IEC Geneva, Switzerland. www.iec.ch"

TABLE 1.2
Time/Current Zones for AC 15–100 Hz for Hand to Feet Pathway (Summary of Zones in Figure 1.3)

Zones	Boundaries	Physiological Effects
AC-1	Up to 0.5 mA curve a	Perception possible but usually no "startled reaction"
AC-2	0.5 mA up to curve b	Perception and involuntary muscular contractions likely but usually no harmful electrical physiological effects
AC-3	Curve b and above	Strong involuntary muscular contractions. Difficulty inbreathing. Reversible disturbances of heart function. Immobilization may occur. Effects increasing with current magnitude. Usually no organic damage expected.
AC-4[a]	Above curve c_1	Pathophysiological effects may occur such as cardiac arrest, breathing arrest, and burns or other cellular damage. Probability of ventricular fibrillation increasing with current magnitude and time.
	c_1–c_2	AC-4.1 Probability of ventricular fibrillation increasing up to 5%.
	c_2–c_3	AC-4.2 Probability of ventricular fibrillation up to about 50%.
	Beyond curve c_3	AC-4.3 Probability of ventricular fibrillation above 50%.

[a] For durations of current flow below 200 ms, ventricular fibrillation is only initiated within the vulnerable period if relevant thresholds are surpassed. With regard to ventricular fibrillation, this figure relates to the effects of current which flows in the path left hand to feet. For other current paths, the heart current factor has to be considered.

Although there is a large discrepancy in the value of the overall body resistance, measurements and experience have shown that a suitable value for this resistance is 1,000 Ω between extremities (hand to hand, hand to both feet, or from one foot to the other). This is the value adopted by the majority of standards and it is the one we shall use throughout this book.

2 Resistance to Ground

2.1 RESISTANCE TO GROUND OF A HEMISPHERE

The hemisphere is the simplest geometrical shape of an earth electrode whose resistance to ground can be easily calculated. If we assume that the region around the hemisphere is divided into concentric hemispherical shells of equal thickness dx (Figure 2.1) and that the earth resistivity is uniform, it is apparent that the shell nearest to the electrode has the greatest resistance since it has the smallest area normal to the current flow. Each successive shell has a larger area and hence a smaller resistance. The total resistance between the electrode and ground can be determined as follows. Let a be the radius of the hemisphere, I the current flowing to ground, and ρ (Ω·m) the earth resistivity. The resistance of a shell of radius x and thickness dx is

$$dR = \rho dx / 2\pi x^2$$

and the resistance between the surface of the electrode and a point at a radial distance r from its center is

$$R = \int_a^r dR = \frac{\rho}{2\pi}\left(\frac{1}{a} - \frac{1}{r}\right) \tag{2.1}$$

If r becomes infinite the absolute resistance to ground of the hemisphere is

$$R_g = \frac{\rho}{2\pi a} \text{ ohms} \tag{2.2}$$

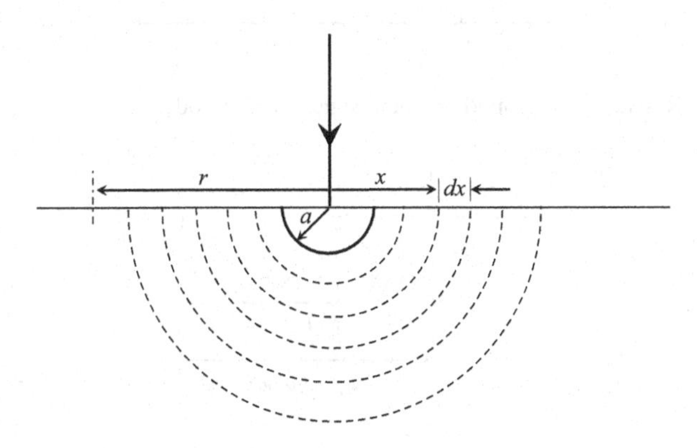

FIGURE 2.1 Hemispherical ground electrode.

Figure 2.2 shows the variation of the resistance to ground as given by Eq. (2.1). From this equation it is apparent that 90% of the absolute resistance of the electrode to ground lies in the region around the electrode whose radius is ten times the radius of the electrode itself.

The potential difference between the electrode and a point at a distance r from its center is given by

$$V_{ar} = IR = \frac{I\rho}{2\pi a} - \frac{I\rho}{2\pi r} \tag{2.3}$$

Since the absolute potential at any point is the potential at that point with respect to a point at infinity (zero potential), the absolute potential of the electrode is (Figure 2.3)

$$V_a = \frac{I\rho}{2\pi a}$$

And the absolute potential at a distance r is

$$V_r = \frac{I\rho}{2\pi r} \tag{2.4}$$

Figure 2.4 gives the potential distribution around the electrode.

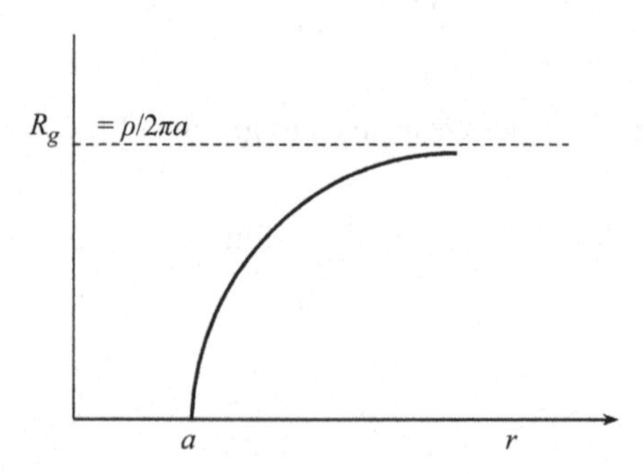

FIGURE 2.2 Resistance to ground of a hemispherical electrode.

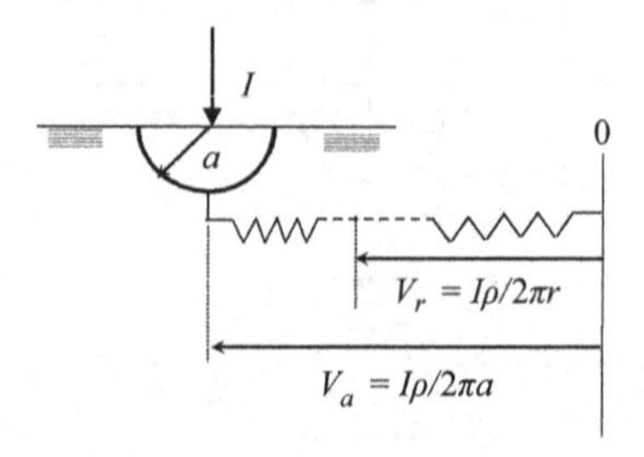

FIGURE 2.3 Absolute potential at the electrode (V_a) and at a distance r (V_r).

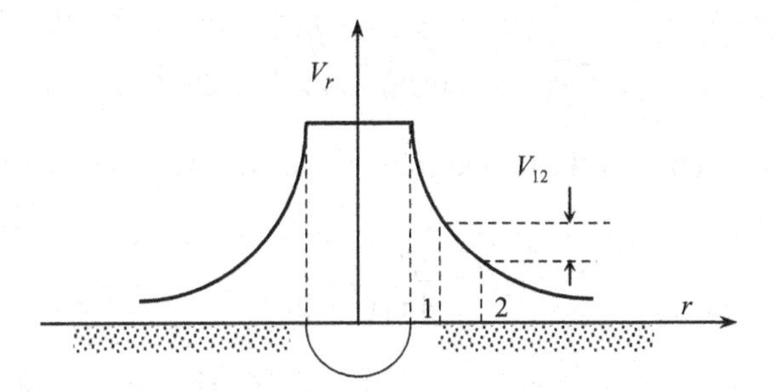

FIGURE 2.4 Potential distribution around a hemispherical electrode.

In order to determine the influence of the burial depth, we shall assume that the conductor which carries the current to the electrode is insulated from ground. If a spherical electrode of radius a is buried in a medium of infinite extent and resistivity ρ Ω·m (Figure 2.5a), it is apparent that in this case

$$R_g = \frac{\rho}{4\pi a}; \quad V_r = \frac{\rho I}{4\pi r}; \quad V_{max} = \frac{\rho I}{4\rho a} \tag{2.5}$$

All these values are one half the corresponding ones for a hemispherical electrode. If the electrode is buried at a depth h meters below the surface of the ground as shown in Figure 2.5b, we can determine its resistance and the potential and field distribution at the earth's surface as follows.

The direction of the current at the earth's surface must be tangential to the surface and this boundary condition is satisfied if we assume that the presence of an image of the buried electrode at a vertical distance $2h$ from its center, i.e., at a distance h above the earth's surface as shown in Figure 2.5b and that the current entering the image electrode is equal to that entering the real electrode. If we assume that $2h >> a$ the potential at the surface of any one of the electrodes is

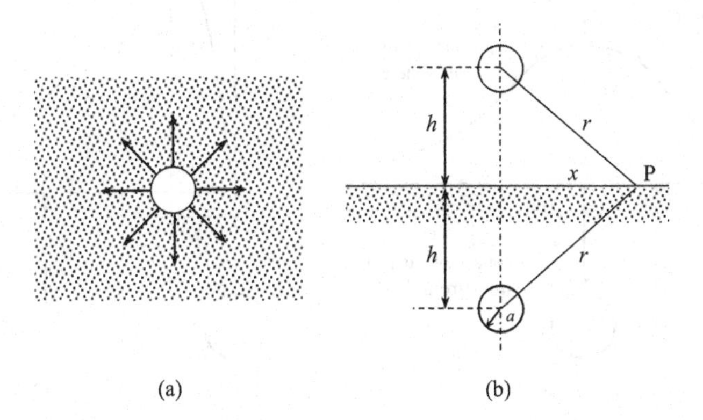

FIGURE 2.5 (a) Electrode buried in infinite medium. (b) Electrode buried at a depth h from the surface.

$$V_o = \frac{I\rho}{4\pi a} + \frac{I\rho}{4\pi(2h)} = \frac{I\rho}{4\pi a}\left[1 + \frac{a}{2h}\right] \tag{2.6}$$

The resistance to ground of the buried sphere is therefore

$$\begin{aligned} R_g = V_o/I &= \frac{\rho}{4\pi a}\left[1 + \frac{a}{2h}\right] \\ &= \frac{\rho}{2\pi a}\left\{\frac{1}{2}\left(1 + \frac{a}{2h}\right)\right\} \end{aligned} \tag{2.7}$$

and the potential of the point P on the ground surface is

$$\begin{aligned} V_x &= \frac{I\rho}{4\pi}\left(\frac{1}{r} + \frac{1}{r}\right) = \frac{I\rho}{2\pi r} \\ &= \frac{I\rho}{2\pi a(x^2 + h^2)^{1/2}} \end{aligned} \tag{2.8}$$

The maximum value of this potential (at $x = 0$) is

$$V_{max} = \frac{I\rho}{2\pi h} = \frac{I\rho}{2\pi a}\left(\frac{a}{h}\right) \tag{2.9}$$

Figure 2.6a shows the potential distribution on the surface of the ground surrounding the buried electrode. It is evident that burying the electrode decreases the potential difference which appears between any two points on the earth's surface.

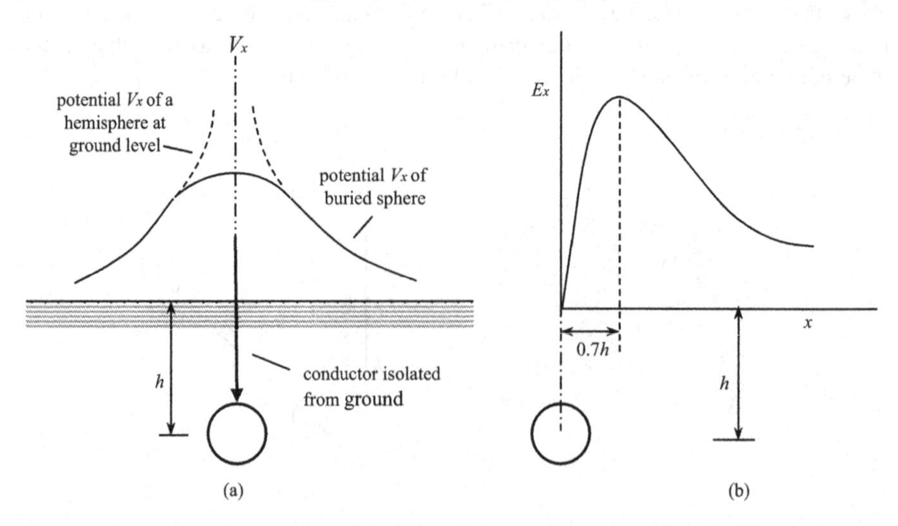

FIGURE 2.6 (a) Potential and (b) electric field distribution on surface of ground around buried electrode.

This is the potential difference which appears between the feet of a person walking in the region around the electrode during the passage of a fault current to ground and is referred to as the *step voltage* which will be discussed in detail in Chapter 4.

The potential difference between anybody connected to the ground electrode and a point on the ground surface immediately above the electrode is

$$V_o - V_{\max} = \frac{\rho I}{4\pi}\left(\frac{1}{a} - \frac{3}{2h}\right) \tag{2.10}$$

This potential difference is referred to as the *touch voltage*, and increases with increasing burial depth.

The electric field at the point P is, from Eq.(2.8),

$$E_x = -\frac{dV}{dx} = \frac{I\rho x}{2\pi(x^2 + h^2)^{3/2}} \tag{2.11}$$

It is zero at a point vertically above the electrode ($x = 0$) and has a maximum value at a distance $x = 0.7h$:

$$E_{\max} = \frac{I\rho}{2\pi(1.6h)^2} \tag{2.12}$$

Figure 2.6b shows the field distribution on the ground surface surrounding the buried electrode.

2.2 MEASUREMENT OF ELECTRODE RESISTANCE TO GROUND

The most reliable and most frequently used method for measuring the resistance to ground of an earth electrode is the so-called *fall-of-potential* method. In this method (Figure 2.7), E represents the earth electrode and C and P are auxiliary electrodes. If a current I flows between C and E and the potential difference between E and P is V, then (subject to the conditions stated below) the ratio V/I gives the required resistance to ground of the electrode E. The source of the current I is the supply S which generates a constant voltage which should be alternating to avoid any electrolytic action and its frequency should be higher than the power frequency, between 70 and 80 Hz, to facilitate the elimination of power-frequency stray currents.

If a series of measurements are made with different values of the distance H a curve similar to that shown in Figure 2.7 is obtained. That part of the curve that is quasi horizontal is the true resistance of the electrode E. As will be shown below the extent of this horizontal section depends on the distance D between electrodes E and C, i.e., on the degree of overlap of the resistance area (see Section 3.6) of these electrodes.

Assume that the electrode E is replaced by an equivalent hemispherical electrode of radius a (see Section 3.1). The current electrode C and the potential electrode P are at distances D and H, respectively, from the center of the hemisphere. Suppose that

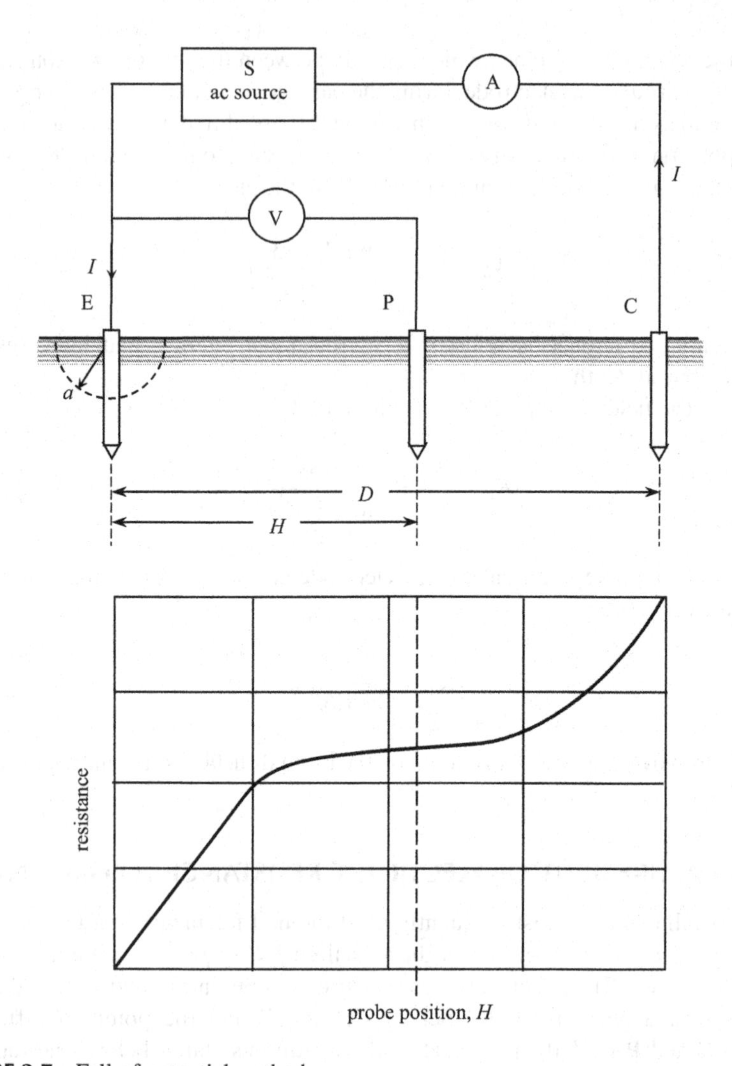

FIGURE 2.7 Fall-of-potential method.

the current I enters at E and leaves at C. With reference to the previous section, it is possible to express the potential at different points as follows:

Absolute potential at E due to current entering:

$$I\rho/2\pi a$$

Absolute potential at E due to current leaving at C:

$$-I\rho/2\pi(\mathrm{D}-a)=-I\rho/2\pi D(D>>a)$$

The absolute potential at point E is therefore

$$V_E=\frac{I\rho}{2\pi}\left(\frac{1}{a}-\frac{1}{D}\right)$$

Absolute potential at P due to current entering at E:

$$I\rho/2\pi H$$

Absolute potential at P due to current leaving at C:

$$-I\rho/2\pi(D-H)$$

The absolute potential at point P is therefore

$$V_P = \frac{I\rho}{2\pi}\left(\frac{1}{H} - \frac{1}{D-H}\right)$$

and the potential difference between points E and P is

$$V_{EP} = V_E - V_P = V$$

$$= \frac{I\rho}{2\pi}\left(\frac{1}{a} - \frac{1}{D} - \frac{1}{H} + \frac{1}{D-H}\right)$$

If we let

$$D/a = c;\ H/a = p$$

the measured resistance between E and P is given by

$$R_{EP} = V/I = \frac{\rho}{2\pi a}\left(1 - \frac{1}{c} - \frac{1}{p} + \frac{1}{c-p}\right)$$

Since the actual resistance of a hemispherical electrode is $\rho/2\pi a$ then the ratio between the measured resistance and the actual resistance is

$$\frac{\text{Measured resistance}}{\text{Actual resistance}} = 1 - \left(\frac{1}{c} + \frac{1}{p} - \frac{1}{c-p}\right) \tag{2.13}$$

The quantity between brackets represents the fractional error which results from the choice of the distances *D* and *H*. The condition for zero error is

$$\frac{1}{c} + \frac{1}{p} - \frac{1}{c-p} = 0$$

$$p^2 + cp - c^2 = 0$$

the positive root of p is

$$p = \tfrac{1}{2}c(\sqrt{5} - 1) = 0.618c$$

that is,

$$H = 0.618D \tag{2.14}$$

This result shows that whatever the distance D between the electrode E and C it is possible to obtain the resistance of electrode E to ground when the distance H between it and the potential electrode P is 61.8% of D.

Figure 2.8 shows a graphical representation of Eq. (2.13). The fractional error in the resistance of electrode E is plotted as a function of the ratio $p = H/a$ for different values of the ratio $c = D/a$. It is apparent that the larger the value of c the smaller the error in the actual value of the resistance of electrode E arising from any error in setting the distance $H = 0.618D$. It has been agreed that the error in the actual resistance should be within ±2%.

Figure 2.9 shows the relationship between the radius a of the hemisphere equivalent to the ground electrode and the distances D and H such that the error lies within ±2%.

When the earthing system consists of a number of electrodes or of a ground network the auxiliary electrodes must be placed outside the area of the earthing system. The potential electrode P should be located at a distance not less than 5 times

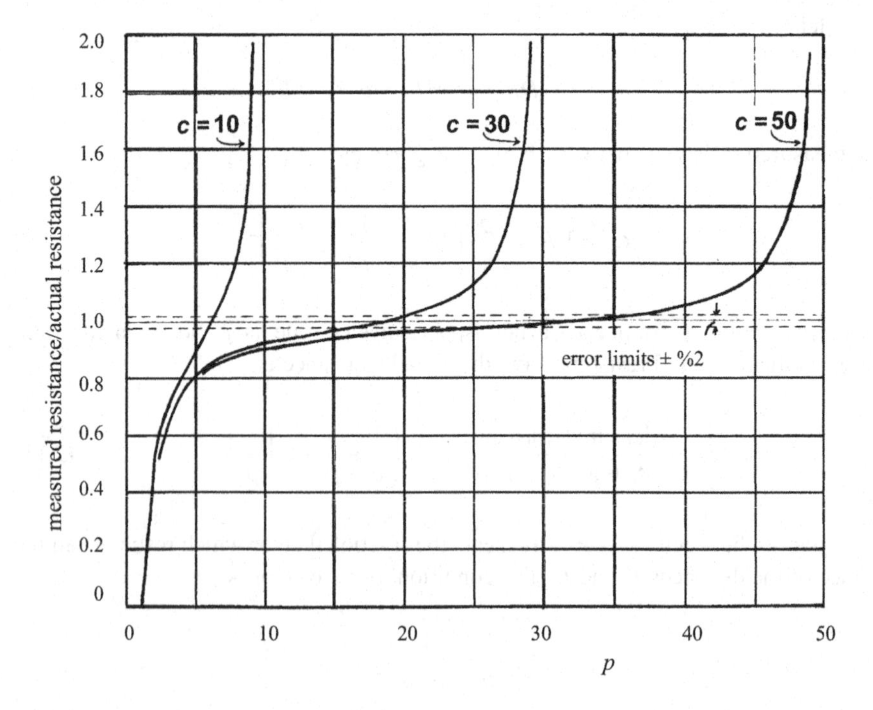

FIGURE 2.8 Graphical representation of Eq. (2.13).

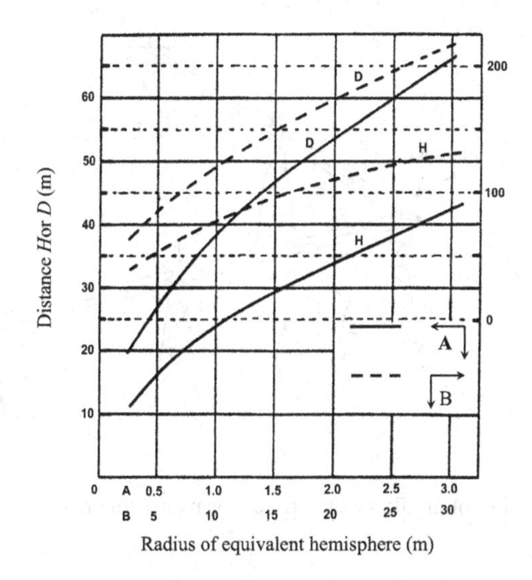

FIGURE 2.9 Separation of auxiliary electrodes to keep measuring error within ±2%. (Curves adapted from Ref. [1].)

the longest radial distance of the area and the current electrode C at a distance not less than 30m from P. In this case the distances D and H are measured from the electrical center of gravity of the grounding system as shown in Figure 2.10. In this case

$$H' = 0.62D'$$
$$(H + X) = 0.62(D + X)$$

$$H = 0.62D - 0.38X \tag{2.15}$$

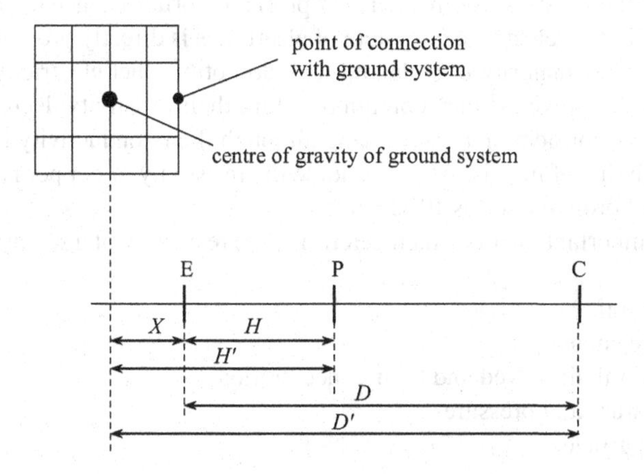

FIGURE 2.10 Measurement of the resistance to ground of an extended grounding system.

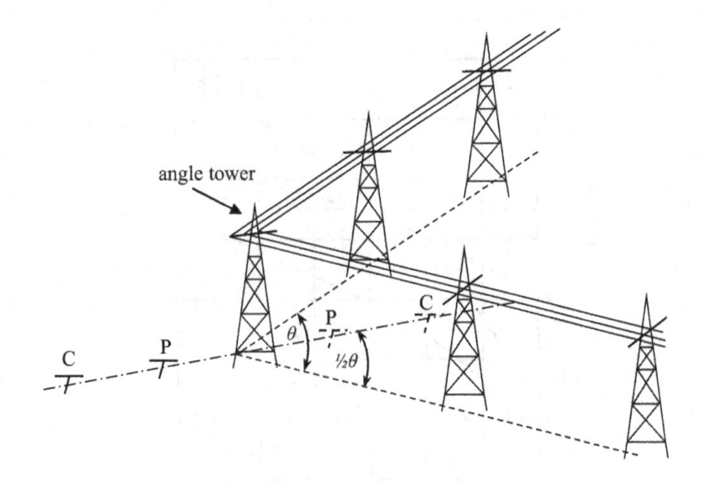

FIGURE 2.11 Positioning of auxiliary electrodes for measuring resistance to ground of an angle tower.

The approximate value of *X* may be determined from the equation

$$X = \sqrt{A/\pi} \tag{2.16}$$

where *A* is the area which encloses the ground electrodes.

When measuring the resistance to ground of any overhead transmission line tower, the auxiliary electrodes should be placed in a direction perpendicular to the direction of the line in order to avoid any error resulting from induction effects. If the tower is an angle tower (Figure 2.11) the auxiliary electrodes should lie along the line bisecting the angle between the two line directions.

2.3 SOIL RESISTIVITY

Knowledge of the resistivity of the soil is of prime importance since the resistance to ground of any buried electrode or system of electrodes is directly proportional to the soil resistivity. The majority of soils and rock do not conduct electricity if they are completely dry. However, if they contain moisture their resistivity decreases greatly and they may be considered as conducting although their conductivity is very small compared with that of metals. As an example the resistivity of copper is 0.017 μΩ·m whereas that of ordinary soil is 100 Ω·m.[1]

The most important factors which determine the resistivity of a soil are as follows:

1. Type of soil
2. Moisture content
3. Types of salt dissolved and their concentration
4. Temperature and pressure
5. Size of particles

[1] The unit of resistivity in ohm-meter represents the resistance of a cube of side 1 m1 Ω·m= 100 Ω·cm

The following table gives the approximate range for the resistivity of different types of soils:

Sandy clay (mixture of clay, sand, and ash)	5–50 Ω·m
Loam	8–50
Mixture of sandy clay and stone	40–250
Sand and gravel	60–100
Sandstone and slate	10–500
Crystalline rock	200–10,000
Concrete (1 part cement + 3 parts sand)	50–300
Concrete (1 part cement + 5 parts gravel)	100–8,000

One of the most important of the factors which influence the soil resistivity is the quantity of water retained in the soil, i.e., its moisture content, and the resistivity of this water itself and thus the type and concentration of salts dissolved in it. The moisture content of any soil changes with changes in the weather, in the seasons, in the nature of the subsoil, and in the depth of the water table. Except in deserts, it is rare for a soil to be completely dry; however, it is also rare for the moisture content to exceed 40%. In general the moisture content varies from about 10% in the dry seasons to about 35% in the rainy seasons.

Measurements indicate that the value of the soil resistivity does not change much if the moisture content exceeds 20%, but changes considerably as the moisture content drops below that value (Figure 2.12). For example, we find that for a moisture content of 10% the resistivity is 30 times that for a moisture content of 20%. Because of such large variations, measurements of soil resistivity for grounding purposes should be carried out in the dry season in order to represent the worst possible conditions as far as resistance to ground of the earthing system is concerned.

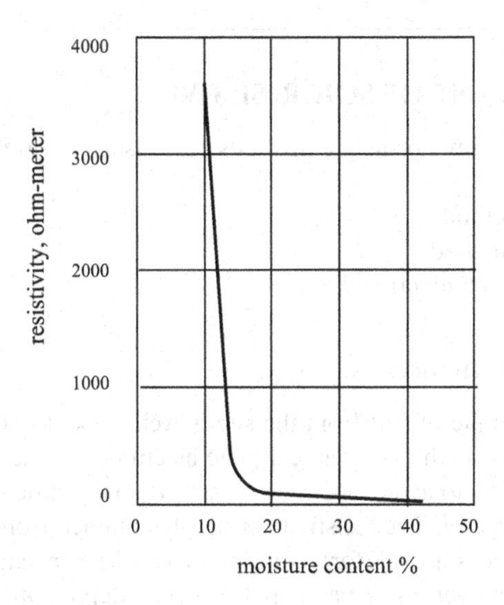

FIGURE 2.12 Variation of the resistivity of clay soil with moisture content.

There is a considerable increase in soil resistivity as the temperature drops below the freezing point. For example the resistivity of sandy clay with about 18% moisture content rises from about 50 Ω·m at 20°C to about 360 Ω·m at – 15°C. This must be taken into consideration when installing ground electrodes in cold regions; the increase in resistivity can be minimized by burying electrodes below the frost line.

It should be pointed out here that in general the resistivity of a soil is associated with its degree of corrosiveness. This is because resistivity is determined mainly by moisture content as well as type and concentration of dissolved salts which such moisture always contains. The following is a rough guide to the correlation between resistivity and corrosiveness:

Resistivity (Ω·m)	Degree of Corrosiveness
<5	Highly corrosive
5–10	Corrosive
10–20	Moderately corrosive
20–100	Mildly corrosive
>100	Slightly

When measurement of the soil resistivity at a certain site is difficult the following values may be used as a guide:

Type of Soil	Resistivity (Ω·m)
Moist organic soil	10
Moist soil	100
Dry soil	1,000
Rocky soil	10,000

2.4 MEASUREMENT OF SOIL RESISTIVITY

There are essentially three principal methods of measuring soil resistivity:

- Simple box method
- The Wenner method
- Electromagnetic induction device

2.4.1 Simple Box Method

In this method a sample of soil from the site is well packed into a rectangular box of insulating material with two identical plate electrodes at each end (Figure 2.13). An AC source (not DC to avoid electrode polarization) is applied between the plates and the current measured. The resistivity is simply obtained from Ohm's law.

This simple method is suitable for testing in situ if it is known that the soil consistency at the site is uniform over a large area and sufficient depth. Otherwise it is a cheap

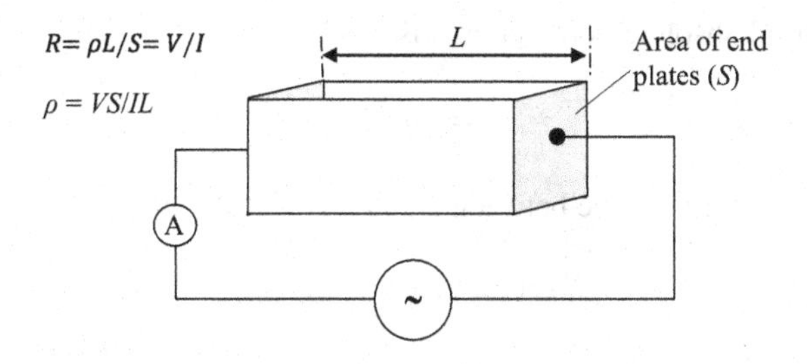

FIGURE 2.13 Simple box for measurement of soil resistivity.

and simple laboratory method to determine the resistivity of different types of soil as well as the effect which moisture, temperature, and additives (such as salts) have on soil resistivity.

2.4.2 Wenner Method

This is the most common method of measuring the soil resistivity (Figure 2.14). In this method four electrodes are driven into the ground along a straight line at equal intervals. A known current I passes between electrodes 1 and 4 and the potential difference V is measured between electrodes 2 and 3. Assuming that the probe length is small compared to electrode separation (which is the usual case) we have that

Absolute potential at 2 due to current entering at 1:

$$I\rho/2\pi L$$

Absolute potential at 2 due to current leaving at 4:

$$-I\rho/4\pi L$$

The total absolute potential at 2 is therefore

$$V_2 = I\rho/4\pi L$$

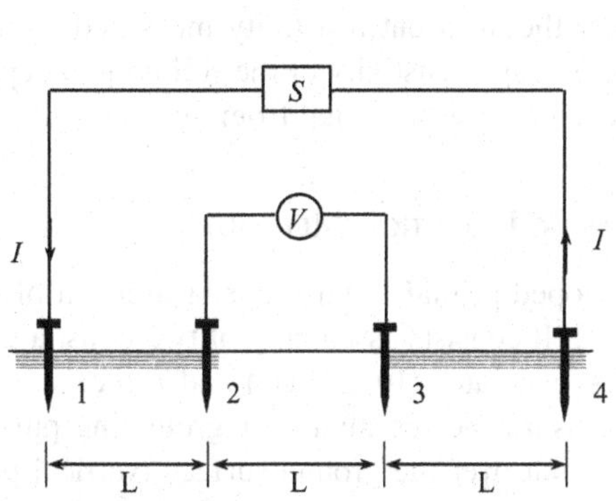

FIGURE 2.14 Wenner method of measuring soil resistivity.

Similarly the total absolute potential at 3 is

$$V_3 = -I\rho/4\pi L$$

Hence the potential difference between 2 and 3 is

$$V_{23} = V_2 - V_3 = V = I\rho/2\pi L$$

and the soil resistivity is given by

$$\rho = (V/I)2\pi L = 2\pi LR \tag{2.17}$$

where ρ is the actual soil resistivity if the soil is homogeneous. However, since the resistivity of a soil usually varies with depth (top-soil layer, subsoil layer), the measured resistivity ρ represents the *apparent resistivity* of the soil at the measuring site. By increasing the probe spacing L, the current penetrates a deeper and longer distance and gives a better estimate of the resistivity if the latter varies appreciably with depth.

If the length of the probe l is not small compared with the probe separation L, then the apparent resistivity is given by

$$\rho = \frac{4\pi LR}{1 + \dfrac{2L}{\sqrt{L^2 + 4l^2}} - \dfrac{L}{\sqrt{L^2 + l^2}}} \tag{2.18}$$

If $L \gg l$ then $\rho = 2\pi\ LR$.

When measuring the soil resistivity at a site it is recommended to carry out several measurements with different values of the distance L between the electrodes to obtain an average value for the resistivity. Any change in the resistivity with changes in L is an indication that the soil is not homogeneous and in particular that the resistivity changes with depth. This is because the greater the distance between the electrodes the greater the depth to which the current penetrates. In such cases it is a common practice to assume that the apparent resistivity measured with the electrodes at a distance L apart is the average resistivity of the soil up to a depth L. Although such an assumption is not accurate it is accepted from the practical point of view.

2.4.3 Electromagnetic Induction Method

This method was developed primarily for the measurement of soil conductivity for geophysical prospecting. It consists essentially of two coils: a transmitter coil T fed by an audio frequency generator (10–20kHz) and a receiver coil R at a distance s away from T. To measure the resistivity for grounding purposes both coils are placed with their plane flat over the ground surface (vertical polarization position) as shown in Figure 2.15. The varying magnetic field produced by the transmission coil T induces eddy currents in the soil beneath. These currents flow in closed loops

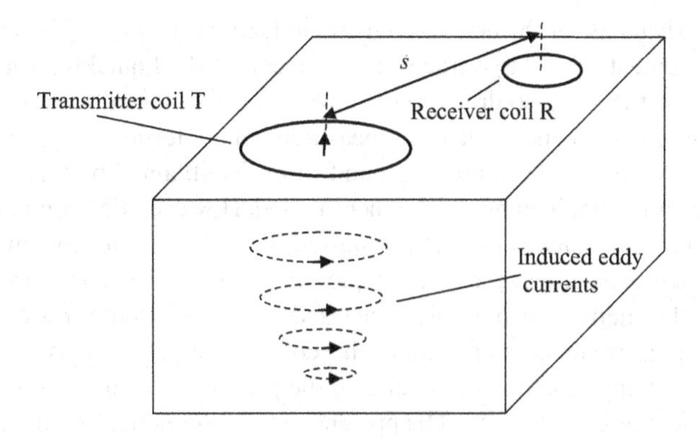

FIGURE 2.15 Principle of electromagnetic induction for measuring soil conductivity (resistivity).

perpendicular to the magnetic field. The magnitude of the currents is proportional to the magnitude of the field, to the rate of change of this field and to the area of the loop, and inversely proportional to the resistivity of the medium in which they flow. The eddy current loops generate their own magnetic field and the receiver coil senses both this field and the field of the primary coil at the receiver location.

Provided that certain design criteria are met[2] the ratio of the secondary to the primary fields is linearly proportional to the apparent soil conductivity and can be expressed as

$$\frac{1}{\rho} = \sigma = \frac{4}{\omega \mu_0 s^2}\left(\frac{H_s}{H_p}\right) \tag{2.19}$$

where

ρ = soil resistivity (Ω·m)
σ = soil conductivity (S/m)
$\omega = 2\pi f$
s = intercoil spacing (m)
H_s = secondary field at receiver loop
H_p = primary field at receiver loop

This linear relationship is valid in the range 1–1,000 Ω·m.

It should be mentioned that measurements can be made with both coils placed either flat on the ground (vertical polarization) or with their planes perpendicular to the ground (horizontal polarization). In the first position the depth of penetration is about twice greater than in the second position and should be used when measuring ρ for grounding purposes. However, measurements made with horizontal polarization provide information on the resistivity of the soil layer near the surface.

[2] J.D.McNeill, *Electromagnetic Terrain Conductivity Measurements at Low Induction Numbers*, Technical Note, Geonics Ltd., 1980.

The great advantage of the electromagnetic induction method (EM) is that it is fast so that measurements over large tracts can be accomplished quickly. Another advantage is that since there are no driven rods it avoids current injection problems (gravel, bedrock, snow, ice, etc.) and is ideal for measurements in terrain where the driving of rods can be very difficult (e.g., rocky ground). Results obtained by this method compare well with those obtained by the Wenner method. However, EM equipment is about 10–20 times more expensive than that required for a Wenner test. Commercial EM measuring instruments are available for exploration depths from 6 to 30m.

Whatever the method of measurement used, a record should always be kept of the date the measurements were made, the condition of the soil (wet or dry), the temperature, and any information regarding the presence (or suspected presence) of bare conductors buried at the site. The presence of such conductors can greatly affect the path of the current flow in the ground and hence the value of the soil resistivity.

2.5 COMPUTER PROGRAMS

The existence of more than one formula for the resistance to ground of most electrode geometries indicates that the formulas are not exact but have nevertheless proved to be very good approximations. In fact they are obtained by mathematical approximations. The choice of the approximation method depends essentially on the geometric configuration of the electrode or electrode system, on the boundary conditions, and on the degree of accuracy required. Numerical methods are best since they can be readily computerized. The two most common of these methods best suited for solving the partial differential equation which governs the system's behavior (e.g., the Laplace equation$\nabla^2 V=0$) are the finite difference method and the finite element (FE) method. Of these the FE method is better suited to systems with irregular geometry, unusual boundary conditions, or nonhomogeneous media.

Essentially the FE method consists of subdividing the region of interest into a number of finite simply shaped elements which could be one-, two-, or three-dimensional. The finer the subdivisions the more closely will the resulting FE structure represent the original configuration. Nodes are the points where elements "connect" and the collection of nodes constitutes the FE mesh. An approximate solution is then developed for each element and the total solution is obtained by assembling the individual solutions such that all boundary conditions are satisfied at all interelement boundaries. Thus the partial differential equation is solved in a piecewise manner. It should be pointed out that the total solution requires the solution of a very large number of simultaneous linear equations. This is such an extremely laborious and time-consuming task that FE methods were only used in simple cases. However, with the advent of powerful and fast computers, FE has become the method of choice for the determination (among its many other applications) of the resistance to ground of a large variety of electrode geometries and configurations (e.g., grounding grids) in both uniform and non-uniform soils. There are a number of commercially available programs such as ETAP, CYMGRD, or CONSOL Multiphysics which use the FE method as an alternative with automatic mesh generation.

It should be pointed out that whatever the method used to predetermine the value of the resistance to ground of a chosen configuration, the actual value can only be obtained by direct measurements after the system has been installed.

3 Earthing Electrodes

3.1 SINGLE DRIVEN ROD

Although the hemispherical electrode is the simplest geometric shape for calculating its resistance to ground, from the practical point of view, the most common type of earthing electrode (especially in distribution networks) is the solid cylindrical rod driven vertically into the ground (Figure 3.1).

A fairly accurate expression for the resistance of a rod electrode is given by Laurent[1]

$$R = \frac{\rho}{2\pi L} \ln \frac{3L}{d} \tag{3.1}$$

where ρ is the soil resistivity, d is the rod diameter, and L is the length of the rod.

Other formulas frequently used are that given by Tagg,[2]

$$R = \frac{\rho}{2\pi L} \ln \frac{4L}{d} \tag{3.2}$$

and that given by Dwight,[3]

$$R = \frac{\rho}{2\pi L} \left(\ln \frac{8L}{d} - 1 \right) \tag{3.3}$$

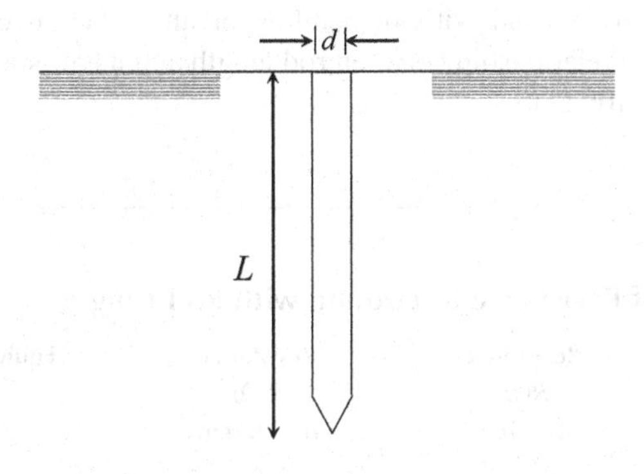

FIGURE 3.1 Driven rod electrode.

[1] P. Laurent, *General fundamentals of electrical grounding techniques*, in IEEE Std 80-1986(R1991): *Guide for safety in AC substation grounding*.
[2] G.E. Tagg, *Earth Resistances*, Newnes, London, 1964.
[3] H.B. Dwight, Calculations of resistances to ground, *Elec. Engr.*, Vol 55, p 1319, 1936.

As a typical example if we assume that $d = 5\,\text{cm}$, $L = 6\,\text{m}$, and $\rho = 100\ \Omega\cdot\text{m}$, we find that:

From Eq.(3.1),$R = 15.60\ \Omega$
From Eq.(3.2),$R = 16.17\ \Omega$
From Eq.(3.3),$R = 15.55\ \Omega$

The differences between these values are of no practical significance especially in view of the inherent inaccuracy in the value of the soil resistivity ρ. In the present text we shall use Eq. (3.1).

By equating Eq.(3.1) with Eq. (2.2) we can obtain the radius a of the equivalent hemispherical electrode, i.e., the hemisphere which has the same resistance to ground as the rod,

$$a = \frac{L}{\ln(3L/d)} \tag{3.4}$$

Because the rod diameter d appears in the logarithmic term in Eq. (3.1), the magnitude of the resistance to ground of a rod electrode does not change significantly with its diameter and we can therefore consider that its resistance is directly proportional to the earth resistivity and inversely proportional to the length of the rod. On the other hand, the resistance to ground of a hemispherical electrode is directly proportional to the earth resistivity and inversely proportional to its radius. If we assume the resistivity increases 100 times then, for a given resistance to ground, the length of a rod would have to be increased 100 times. The radius of the hemisphere would also have to be increased 100 times and hence the excavated volume would increase a million times. This example shows the principal advantage of using driven rods as ground electrodes.

Table 3.1 gives the variation of resistance with rod length for two rod diameters. It is apparent that the rod diameter has no significant effect on the rod resistance to ground. In practice the diameter of the rod is chosen such that the rod can withstand being driven into the ground without bending or any other mechanical damage. Figure 3.2 shows the relationship between rod length and its resistance to ground for $d = 2.5\,\text{cm}$ and $\rho = 100\ \Omega\cdot\text{m}$.

TABLE 3.1
Variation of Resistance to Ground with Rod Length

Length L(m)	Resistance $R(\Omega)$ $d = 5$cm	Resistance $R(\Omega)$ $d = 2.5$cm	Equivalentradius a(m) d =2.5cm
1	$0.61\,\rho$	$0.76\,\rho$	0.21
2	$0.38\,\rho$	$0.44\,\rho$	0.36
4	$0.22\,\rho$	$0.25\,\rho$	0.64
8	0.123ρ	0.137ρ	1.16
16	$0.07\,\rho$	0.075ρ	2.12
32	$0.04\,\rho$	$0.04\,\rho$	3.98

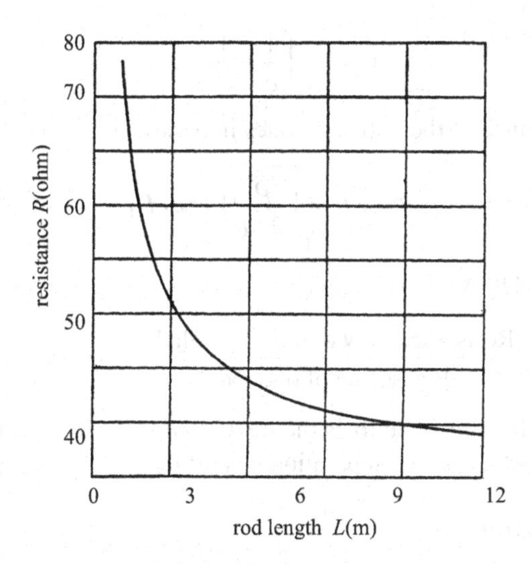

FIGURE 3.2 Relationship between R and L for a driven rod.

3.2 GROUNDING USING MULTIPLE RODS

In many cases it is found that extending the length of a single rod is not enough to attain the required value of resistance to ground especially when the soil resistivity is high. In such cases a number of rods connected in parallel must be used.

When a number of electrodes in parallel are used, the resistance to ground is not reduced by the same number unless the distance between the electrodes is infinite. Since this is not practical, it is necessary to know what effect the distance separating the electrodes has on the reduction of the resistance of a single electrode. In the following we will calculate this reduction for two electrodes and for three electrodes in a straight line.

(a) *Two rods in parallel*

Suppose that we have two identical rods at a distance d apart and that the radius of the equivalent hemisphere is a (Figure 3.3a). Because of symmetry each electrode will carry a current of $\frac{1}{2} I$ to ground. With reference to Section 2.1 we may write the absolute potential of any one of the electrodes as

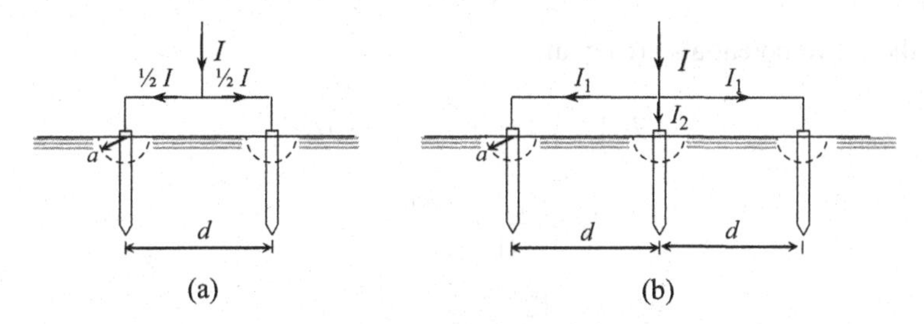

FIGURE 3.3 (a) Two rods in parallel; (b) three rods in parallel.

$$V = \frac{\rho}{2\pi}\left[\frac{1}{a} + \frac{1}{d}\right]\tfrac{1}{2}I$$

and the total resistance of the two electrodes in parallel is

$$R = V/I = \tfrac{1}{2}\frac{\rho}{2\pi a}[1 + (a/d)]$$

If we let $\alpha = a/d$ we may write,

$$\frac{\text{Resistance of two rods in parallel}}{\text{Resistance of one rod}} = \tfrac{1}{2}(1+\alpha) \quad (3.5)$$

The above ratio will be referred to as the *resistance ratio.* It is evident that this ratio does not equal 0.5 except if the separation d is infinite.

(b) *Three rods in parallel*

We shall assume that the three rods are along a straight line as shown in Figure 3.3b. Because of symmetry the currents flowing in the outer electrodes are equal. If I_1 is the value of each of these currents and I_2 is the current in the central electrode then,

$$I = 2I_1 + I_2 \quad (3.6)$$

and the absolute potential of any of the two outer electrodes (assuming that $d >> a$) is

$$V = \frac{I_1\rho}{2\pi a} + \frac{I_2\rho}{2\pi d} + \frac{I_1\rho}{2\pi(2d)}$$

$$= \frac{\rho}{2\pi a}\left[I_1 + \alpha I_2 + \tfrac{1}{2}\alpha I_1\right]$$

$$= \frac{\rho}{2\pi a}\left[I_1(1 + \tfrac{1}{2}\alpha) + \alpha I_2\right]$$

and the potential of the central electrode is

$$V = \frac{\rho}{2\pi a}[I_2 + 2\alpha I_1]$$

Since these two potentials are equal,

$$I_1(1 + \tfrac{1}{2}\alpha - 2\alpha) = I_2(1-\alpha)$$

$$I_2 = \frac{I_1(1 - \tfrac{3}{2}\alpha)}{(1-\alpha)} = kI_1$$

$$k = \frac{(2-3\alpha)}{2(1-\alpha)} \quad (3.7)$$

From Eqs. (3.6) and (3.7) we find that

$$I_1 = \frac{I}{2+k}$$

and the potential V can be written as

$$V = \frac{\rho}{2\pi a}\left[\frac{kI}{2+k} + \frac{2\alpha I}{2+k}\right]$$

$$= \frac{\rho I}{2\pi a}\left[\frac{k+2\alpha}{2+k}\right]$$

Substituting for the value of k we find that the resistance ratio is

$$\frac{\frac{(2-3\alpha)}{2(1-\alpha)}+2\alpha}{2+\frac{(2-3\alpha)}{2(1-\alpha)}} = \frac{2-3\alpha+4\alpha-4\alpha^2}{4-4\alpha+2-3\alpha} = \frac{2+\alpha-4\alpha^2}{6-7\alpha}$$

Table 3.2 shows the resistance ratio for a number of electrode configurations and Figure 3.4 shows a set of curves from which the resistance ratio can be obtained for these configurations. It has been found that the calculated values of this ratio are always between 5% and 20% higher than the measured value and is therefore always on the safe side.

When the number of rods required to obtain a certain value for the resistance to ground increases, the most common configurations used are the following (Figure 3.5):

(a) Hollow square: If the number of rods in each side is n then the total number of rods is $4n - 4$.
(b) Solid square.
(c) Rods distributed around a circle.

In many cases the area available for grounding purposes is limited. In such cases it is important to know the number and configuration of rods which will give the optimum utilization of the available area in order to obtain an effective grounding

TABLE 3.2
Resistance Ratio of Different Rod Configurations

Configuration	Resistance Ratio
Three rods in a straight line	$(2 + \alpha - 4\alpha^2/(6-7\alpha)$
Three rods in equilateral triangle	$(1+2\alpha)/3$
Four rods in a straight line	$(12+16\alpha-21\alpha^2)/(48-40\alpha)$

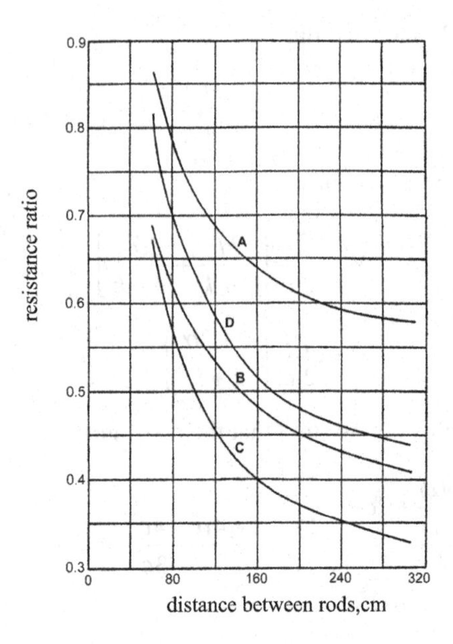

FIGURE 3.4 Relationship between resistance ratio and distance between rods for a number of configurations. A, two rods; B, three rod in a straight line; and C, four rods in a straight line. (Adapted from Ref. [1].)

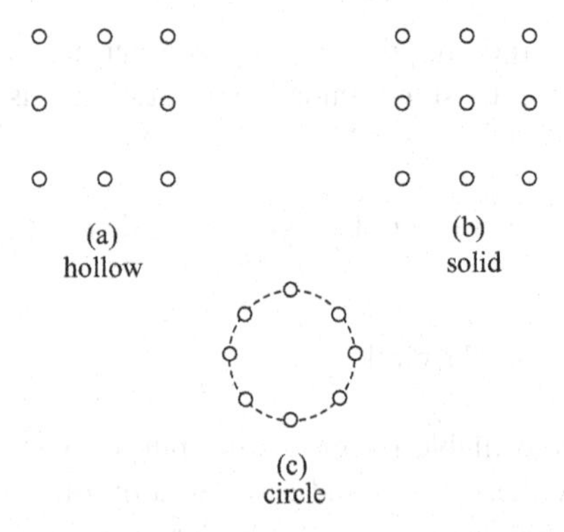

FIGURE 3.5 Rod configurations.

system. Figure 3.6 shows the relationship between the resistance to ground and the number of rods in a hollow square arrangement required to attain this value within different areas. Figure 3.6 indicates that there is a lower limit to the value of the resistance to ground which may be attained and also an economic limit to the number of rod which must be used to obtain this resistance. For example, for an area of 36 m^2, we find that the lower limit for the resistance is 6 Ω and that the economic number of rods is 16. If a resistance to ground of less than 6 Ω is required then a larger area

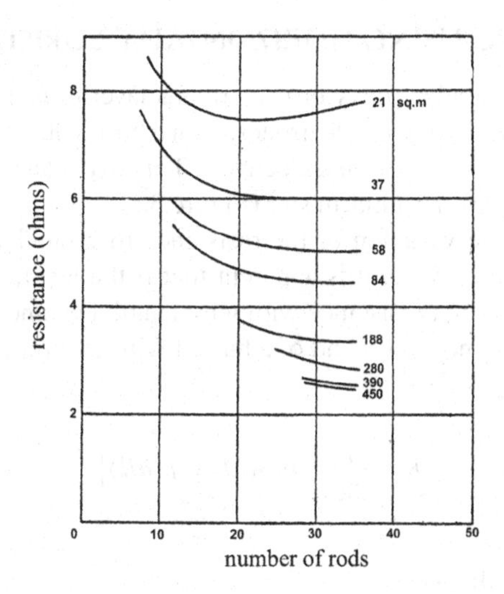

FIGURE 3.6 Resistance to ground of rods arranged in a hollow square of finite area. (Adapted from Ref. [1].) Rod length, 2.4 m; rod diameter, 2.5 cm; $\rho = 100\ \Omega\cdot\text{m}$.

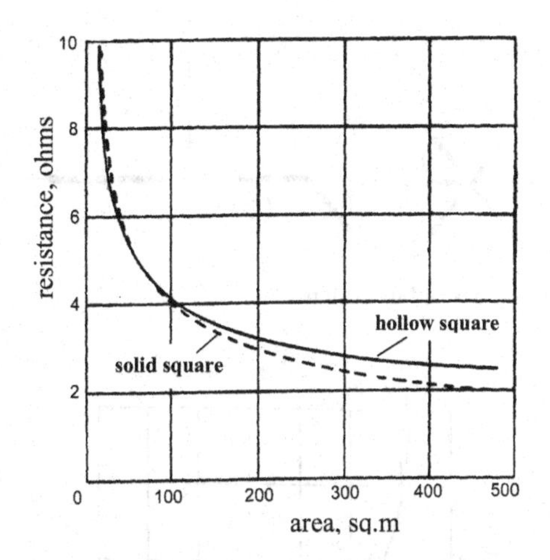

FIGURE 3.7 Minimum resistance obtainable with hollow and solid square rod configurations in a limited area [1]. Rod length, 2.4 m; rod diameter, 2.5 cm; $\rho = 100\ \Omega\cdot\text{m}$.

must be used. If the earth resistivity was 2000 Ω·m instead of 100 Ω·m the lower resistance limit would be 120 Ω (6 × 2000/100) instead of 6 Ω.

Figure 3.7 shows the relationship between area and lower resistance limit for both hollow square and solid square rod configurations. It is evident that if the area is limited then the addition of rods inside a hollow square produces only a very small decrease in resistance value which makes the addition of such rods uneconomical.

3.3 GROUNDING USING HORIZONTALLY BURIED WIRES

If the nature of the ground is rocky or if there is a layer of rock near the surface so that it is difficult to use driven rod electrodes, then a horizontal wire buried at a depth of from 0.5 to 1 m is used as a ground electrode. This wire can have different shapes: straight line, right angle, or multiarm star (Figure 3.8).

Figure 3.9 shows the variation of the resistance to ground with the length of a straight horizontal buried wire. It is apparent that if the length of the wire exceeds 90 m the rate of decrease of resistance with wire length becomes very small.

In general the resistance to ground of a buried wire electrode may be found from the following equation:

$$R = \frac{\rho}{2\pi L}\left[\ln(4l/d) + f(h/l)\right] \tag{3.8}$$

where

L = total length of the wire
l = length of one arm
d = wire diameter
h = depth of burial

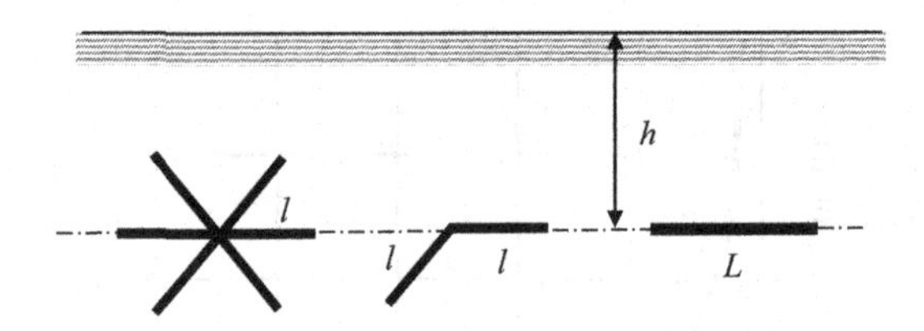

FIGURE 3.8 Common shapes of buried wire.

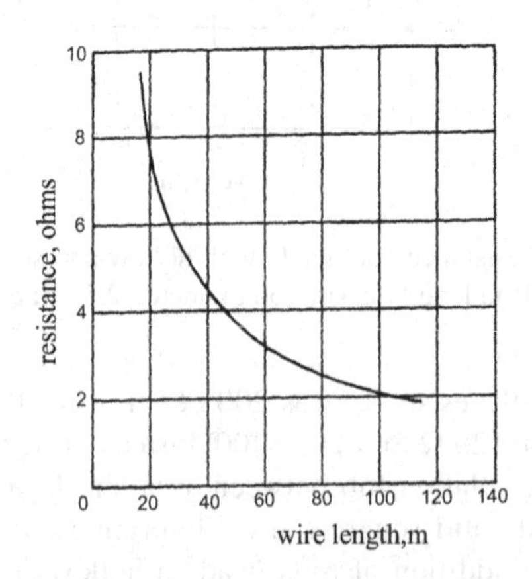

FIGURE 3.9 Variation of resistance to ground with wire length for a horizontal wire buried at a depth of 60 cm (wire diameter, 2.5 cm; $\rho = 100\ \Omega\cdot\text{m}$).

The function $f(h/l)$ is a depth factor whose value depends on the wire configuration. Its value may be obtained from the curves given in Figure 3.10 for six different wire shapes [1]. As an example if $L = 150$ m, $d = 2.5$ cm, $h = 90$ cm, $\rho = 100$ Ω·m, the resistance values to ground of the different wire shapes are as follows:

Straight wire 1.41Ω
Right angle .... 1.44 Ω
3-arm star ....1.50 Ω
4-arm star ....1.62 Ω
6-arm star ....1.94 Ω
8-arm star ....2.33 Ω

The above figures show that for a fixed length of wire the straight wire gives the least resistance. If we now assume that we have a limited area whose diameter is say 60 m, then for the same above values of d, h, and ρ, the resistance, the arm length and the total wire length for the different wire configurations which can be accommodated within the given area are as shown in Table 3.3.

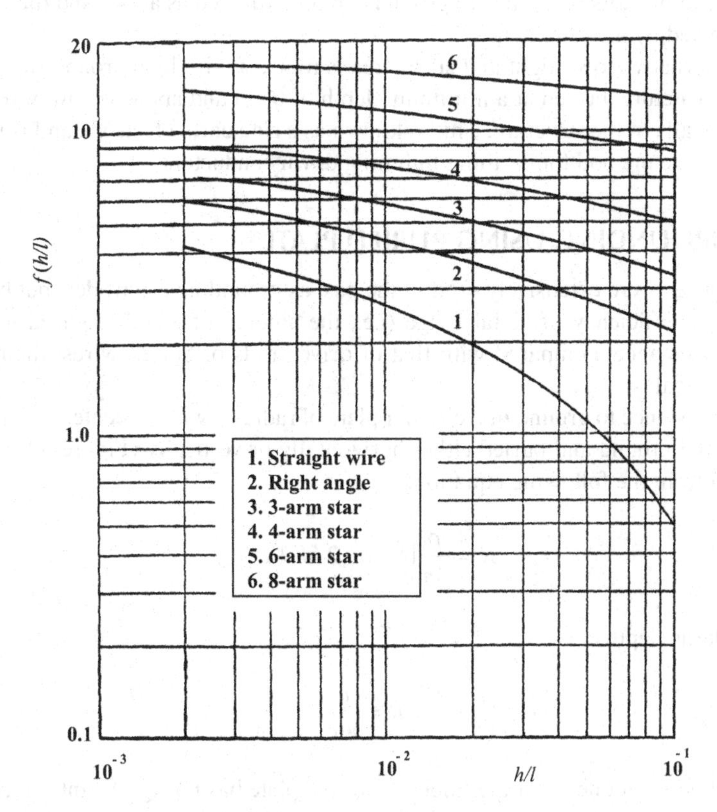

FIGURE 3.10 Depth factor $f(h/l)$ for different configurations of buried wire [1].

TABLE 3.3
Resistance to Ground of Different Wire Shapes

Configuration	*R* (Ω)	*l* (m)	*L* (m)
Straight wire	3.0	60	60
Right angle	2,26	42.4	84.8
3-arm star	2.31	30	90
4-arm star	1.99	30	120
6-arm star	1.63	30	180
8-arm star	1.49	30	240

Area diameter 60 m, d = 2.5 cm, h = 90 cm, ρ = 100 Ω·m.

From this Table 3.3 we see that the resistance of an 8-arm star is half that of a straight wire. However, because a very large increase in wire length is required to achieve only a small decrease in resistance, the common practice is to limit the number of star arms to a maximum of four.

It should be mentioned here that the straight wire could very well be a buried water metal pipe; its resistance to ground can be estimated as above and then preferably checked by measurement.

An alternative to straight and star wires is to use a wire loop around the periphery of a structure, buried at a minimum depth of 0.5m and about 1m away from the external walls of the structure. This is known as a ring earth electrode and is used as an alternative method for earthing lightning down conductors.

3.4 GROUNDING USING BURIED PLATES

Buried plates were extensively used in the past as grounding electrodes, but because of the poor efficiency of metal usage (i.e., the amount of metal required to attain a given resistance) compared with that of driven rods or buried wires, their use is uneconomical.

The resistance to ground of a circular plate of radius a with its center at a distance h below the ground and buried either horizontally or vertically (Figure 3.11) can be obtained from the following equation:

$$R = \frac{\rho}{8\pi}\left[1 + a/(2.5h + a)\right] \tag{3.9}$$

and for large depths,

$$R = \frac{\rho}{8\pi} \tag{3.10}$$

It should be mentioned that the thickness of the plate has no significant effect on its resistance to ground.

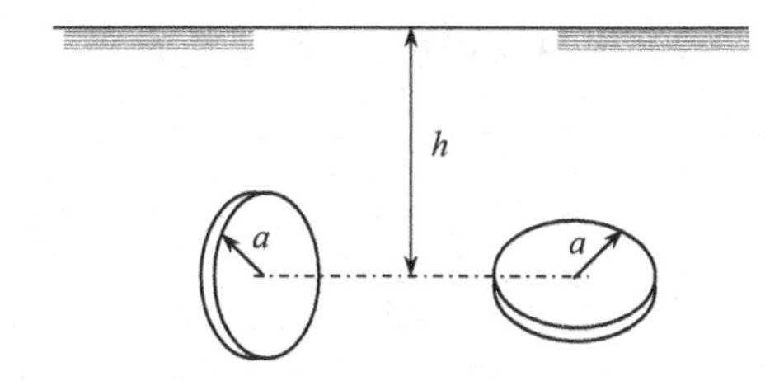

FIGURE 3.11 Buried circular plate.

3.5 WATER PIPES OR STEEL REINFORCEMENT AS GROUNDING ELECTRODES

It is sometimes possible to use the metal pipes of a water network as earthing electrodes. However, before such a decision is taken, the network's resistance to ground must be measured and the electrical continuity of all the joints must be ensured (by means of jumpers). The electrical continuity must be checked after any repair or maintenance work is carried out on the system. If the replacement of metal pipe section by plastic pipes is envisaged or if the electrical continuity cannot be guaranteed, then this method should not be used.

In buildings where steel reinforcement is used in the concrete foundations such as column bases, the reinforcement bars can be used as grounding electrodes. The resistivity of the concrete below the surface of the earth is about 30 Ω·m which is lower than the average resistivity of the soil itself (100 Ω·m). In such cases it is sufficient to make the connections between any main reinforcing rod and the main earth connection at the distribution board.

3.6 THE RESISTANCE AREA

Figure 2.2 shows that as the distance from the grounding electrode increases, the ratio between the value of the resistance up to this distance and the total (absolute) resistance of the electrode increases. It is clear that the rate of this increase is rapid at first but then becomes asymptotic. Theoretically the total resistance to ground can only be attained at infinite distances. For practical purposes, we can say that surrounding any electrode there is a finite region of ground that contains the "major part" of the resistance. There is general agreement that this major part should represent 98% of the total resistance and it is known as the *resistance area*.

As shown in Chapter 2 the resistance between a hemispherical electrode of radius a and a point at a distance r from its center is

$$R = \frac{\rho}{2\pi}\left(\frac{1}{a} - \frac{1}{r}\right)$$

The ratio x of this resistance to the total resistance $\rho/2\pi a$, expressed as a percentage, is

$$x = 100(1 - a/r)$$

so that

$$r = \frac{100a}{100 - x} \tag{3.11}$$

If we assume that the grounding electrode is a rod of length 2.5 m (which is the minimum length permitted) and diameter 2.5 cm, then the radius of the equivalent sphere from Eq.(3.4) is 43.2 cm and the radius of the resistance area is

$$r = 43.2 / (100 - 98) = 21.6\,\text{m}$$

If it is required to obtain an effective electrical separation between two different grounding electrodes then the distance between them should not be less than 20 m. If the distance is less then there is an overlap of their respective resistance areas. Electrically this means that there is a mutual resistance between the two grounding points (Figure 3.12a). The effect of this resistance may be clarified from Figure 3.12b; when a short circuit occurs between one phase and the body of the transformer, the potential of the point E rises to a value which depends on the magnitude of the resistance to ground at E and that of the short circuit current. The potential of the point S, and hence of the body grounded through S, rises to a value determined by the value of the mutual resistance R_m between points S and E. The closer the points the higher the mutual resistance.

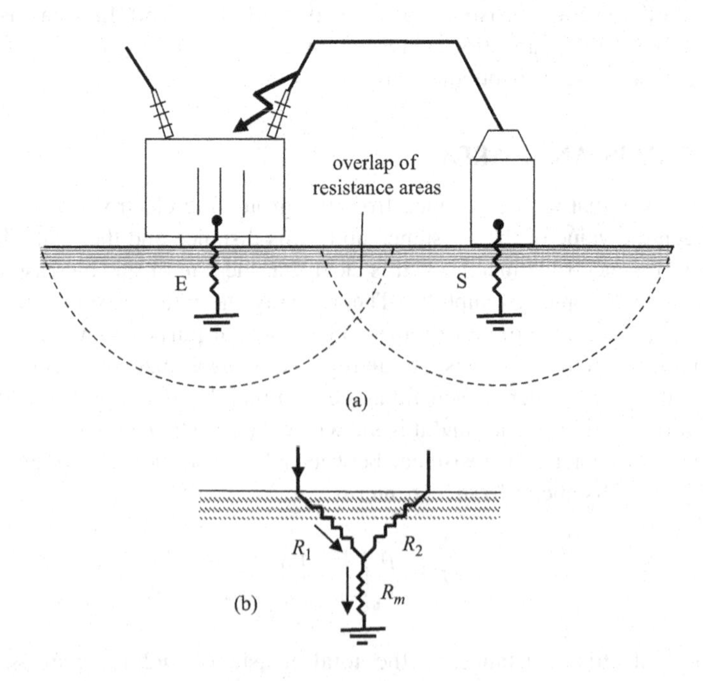

FIGURE 3.12 Overlap of resistance areas.

4 Step, Touch, and Transfer Voltages

4.1 STEP VOLTAGE

If a man is standing with his feet apart or walking in an area in the vicinity of a ground electrode (in particular a substation ground or a short-circuit testing facility ground) and a short-circuit current I flows to ground then a voltage, known as the step voltage, will appear across his feet. Suppose that the man is at a distance x from the electrode and that the distance between his two feet is s. If the ground electrode is represented by an equivalent hemisphere of radius a, then the absolute potential at a distance x from its center is (from Eq.2.4)

$$V_x = \frac{I\rho}{2\pi x}$$

and the potential difference appearing between the man's feet is therefore

$$V_{step} = \frac{I\rho}{2\pi}\left(\frac{1}{x} - \frac{1}{x+s}\right) = \frac{I\rho}{2\pi}\left(\frac{s}{x(x+s)}\right)$$

It is evident that the maximum value of this potential occurs at $x=a$. If we assume that the length of the step s is one meter, the maximum step voltage is

$$V_{step}(\text{max.actual}) = \frac{I\rho}{2\pi}\left(\frac{1}{a(a+1)}\right)$$

If we assume that the resistance of the body is R_b and the resistance between each foot and ground is R_f (Figure 4.1), then the current passing through the body is

$$I_b = V_s / \left(R_b + 2R_f\right)$$

Since the current which passes through the human body must not exceed $0.116/\sqrt{t}$ (see Section 1.3), then the safe step voltage, i.e., the maximum step voltage permitted is given by

$$\begin{aligned} V_{step}(\text{max.permitted}) &= I_b\left(R_b + 2R_f\right) \\ &= (1{,}000 + 2R_f)0.116/\sqrt{t} \end{aligned} \tag{4.1}$$

and it must therefore always be ensured that

$$V_{step}(\text{max.permitted}) \le V_{step}(\text{max.actual})$$

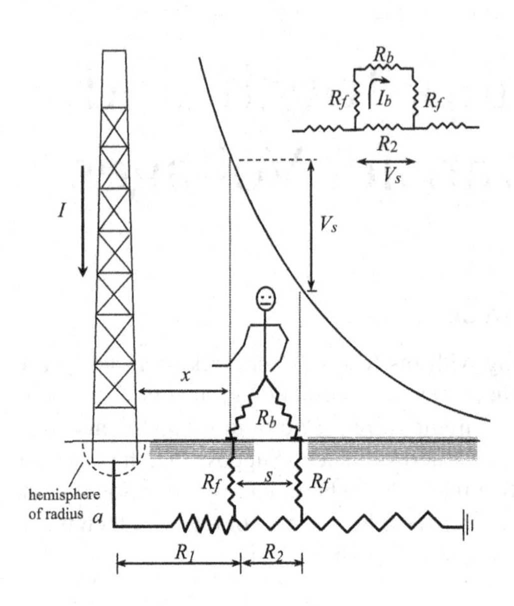

FIGURE 4.1 Step voltage.

4.2 TOUCH VOLTAGE

If a person touches a metal object directly connected to ground at the same instant as a short circuit current is flowing to ground (Figure 4.2), the magnitude of the current which flows between his hand and feet is determined by the so-called touch voltage. Since $I >> I_b$, the touch voltage may be written as

$$V_{touch} = IR_1 = I_b\left(R_b + \tfrac{1}{2}R_f\right)$$

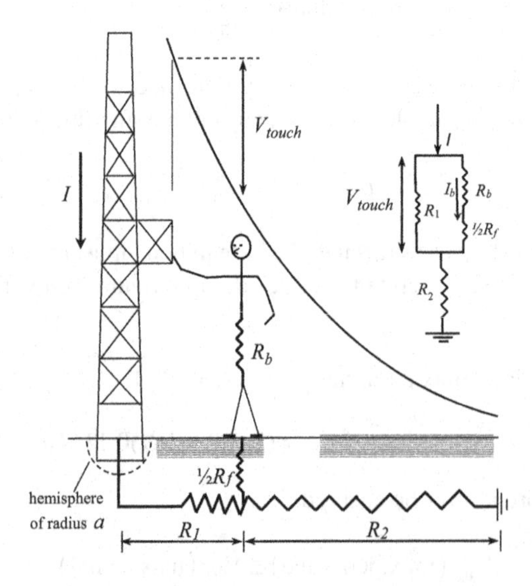

FIGURE 4.2 Touch voltage.

If we assume that the person is at a distance of 1 m from the metal object which he touches then from Eq. (2.1),

$$R_1 = \frac{\rho}{2\pi}\left(\frac{1}{a} - 1\right)$$

setting,

$I_b = 0.116 / \sqrt{t}$ and $R_{b=}$ 1,000 Ω

we find that the safe touch voltage, which is also the maximum permissible touch voltage, is

$$V_{touch}\left(\text{max.permitted}\right) = \left(1{,}000 + \tfrac{1}{2} R_f\right) 0.116/\sqrt{t} \tag{4.2}$$

It is apparent from Eqs. (4.1) and (4.2) that the magnitude of the resistance R_f between foot and ground has a very great influence on the maximum permissible value of both the step and touch voltages. For practical purposes it is possible to regard the foot as a circular electrode whose radius is about 8 cm; the resistance to ground of such an electrode is obtained from Eq.(3.9) for $h = 0$, $a = 0.08$,

$$R_f = \frac{\rho}{4a} = \frac{\rho}{4 \times 0.08} \cong 3\rho$$

where ρ is the resistivity of the soil. Substituting this value of R_f in Eqs. (4.1) and (4.2) we find that maximum permissible step voltage is

$$V_{step}\,(\text{max.permitted}) = \frac{(116 + 0.7\rho)}{\sqrt{t}} \tag{4.3}$$

and that maximum permissible touch voltage is

$$V_{touch}\,(\text{max.permitted}) = \frac{(116 + 0.17\rho)}{\sqrt{t}} \tag{4.4}$$

Equations (4.3) and (4.4) are valid for uniform soil resistivity. In many cases, however, the resistivity is not uniform either because of the nature of the soil or because the ground surface has been covered with a layer of high resistivity (ρ_s) such as crushed rock. In this case these equations are modified as follows:

$$V_{step}\,(\text{max}) = \frac{(116 + 0.7 C_s \rho_s)}{\sqrt{t}} \tag{4.5}$$

$$V_{touch}\,(\text{max}) = \frac{(116 + 0.17 C_s \rho_s)}{\sqrt{t}} \tag{4.6}$$

where C_s = correction factor for derating the surface resistivity ρ_s.

The risk of electrocution is higher for touch voltages than for step voltages as in the former the current traverses the chest region and its path is close to the heart.

It is evident from the last equations that the higher the surface resistivity the higher the permissible step and touch voltages. This is very desirable at large substations where the ground fault current is generally high as well as around lightning ground electrodes. In practice the surface resistivity is increased by covering the surface with a layer of crushed rock such as granite of thickness 10–12 cm. Granite has a very high resistivity when dry and also when wet: dry $1.3 \times 10^6 \Omega\cdot\text{m}$, wet $4.5 \times 10^3 \Omega\cdot\text{m}$. If granite is not available then asphalt may be used as it too has a high wet resistivity $> 10^4 \Omega\cdot\text{m}$ [4].

The sudden change in resistivity at the boundary between the two layers is described by the reflection factor K defined as

$$K = \frac{\rho - \rho_s}{\rho + \rho_s} \tag{4.7}$$

Figure 4.3 shows the relationship between the correction factor C_s and the thickness of the surface layer h_s for different values of K (IEEE Std.80-2000).

Equations (4.5) and (4.6) must be used whenever a surface layer (crushed rock or any other material) has been added or when the soil can be represented by two layers (e.g., top soil + subsoil). It is also possible to obtain an approximate value for C_s from the following formula:

$$C_s = 1 - \frac{0.09\left(1 - \dfrac{\rho}{\rho_s}\right)}{2h_s + 0.09} \tag{4.8}$$

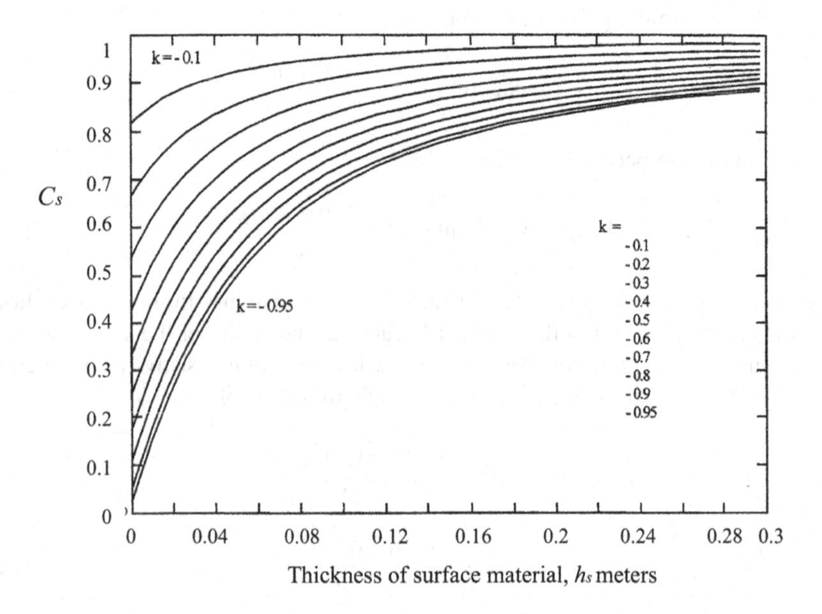

FIGURE 4.3 Relation between correction factor Cs and thickness of surface layer hs. (Reproduced with permission from IEEE Std. 80-2000.)

4.3 TRANSFER TOUCH VOLTAGE

The mutual resistance between two separately earthed systems is, as mentioned in Section 3.6, a source of transfer voltage. In general the transfer touch voltage can be defined as the potential which appears between the grounded metallic parts of two systems which are separately grounded when a fault to ground occurs on one of these systems.[1] This potential can be especially dangerous when one location is a power substation, since under fault conditions the ground potential rise may be of the order of several thousand volts. Two such cases are shown in Figure 4.4.

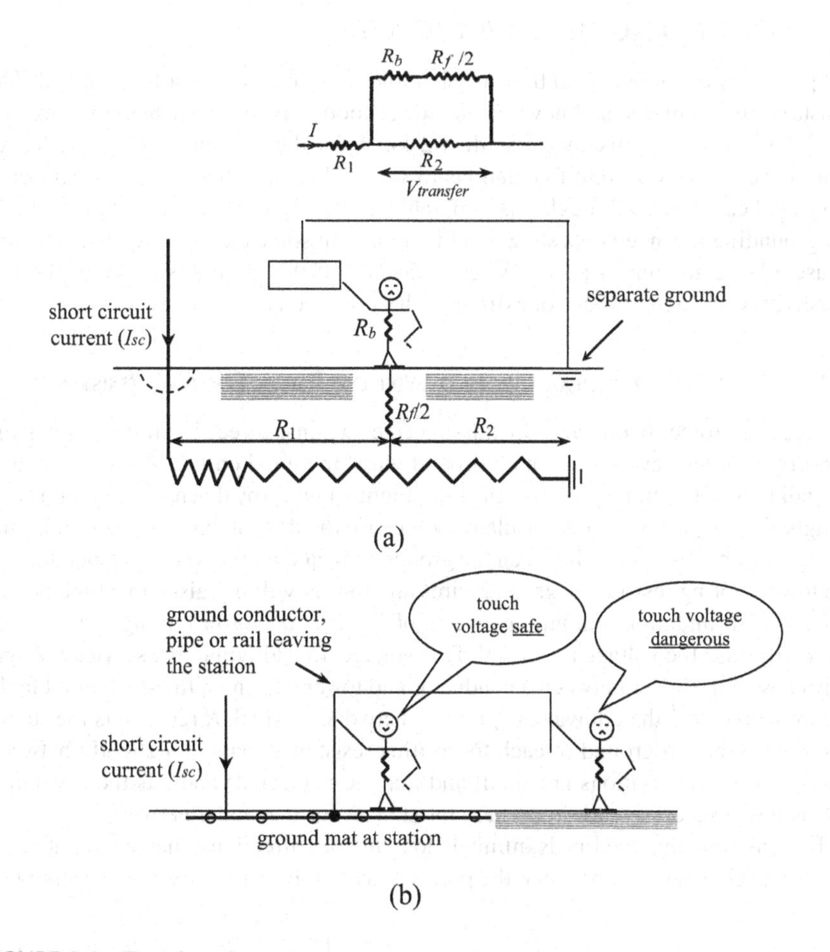

FIGURE 4.4 Transfer voltage.

[1] Separate grounding at the power source and at the utilization (consumer) end is also the source of common-mode noise (voltage and current). Treatment of this topic is beyond the scope of this book and the interested reader may consult any specialized text on the subject, e.g. H.W.Denny, *Grounding for the Control of EMI*, Don White Consultants, Inc.1983, USA.

In Figure 4.4a the absolute potential at the point where the man is standing during the flow of a current I to the earth of a grounded system is $V_x = I\rho/2\pi x = I R_2$. If the man comes in contact with a metallic body which is earthed through a second ground system which is physically separated from the first, then this is the value of the touch voltage he will experience.

In Figure 4.4b the man standing on the properly designed (see Chapter 6) substation ground mat is in no danger under ground fault conditions. However, if a man standing outside the station were to touch a ground conductor or pipe or rail leaving the station he may be in considerable danger.

4.4 GROUNDING OF POWER TOWERS

All power towers of overhead transmission lines must be individually grounded. The resistance to ground, also known as the tower footing resistance, should not exceed 10 Ω for towers at a distance less than 2 km from the line ends and 20 Ω for all other towers. The concrete foundations to which the tower feet are fixed can serve as ground electrodes although it is preferable, in the majority of cases, to use one of the grounding arrangements shown in Figure 4.5. In some cases it may be necessary to use a large amount of buried wires, especially if the ground is rocky or the soil resistivity very high because of extremely dry conditions.

4.4.1 Effect of Overhead Ground Wire on Tower Footing Resistance

In order to protect overhead transmission lines against direct lightning strokes the majority of lines have one or more ground wires mounted on top of the towers and extending over the entire length of the line (Figure 4.6a). Any discharge to ground due to lightning or flashover at an insulator will cause the flow of current to ground. This current will be distributed between the ground wire and all the towers. Depending on the tower footing resistance, ground wires and towers will be raised to a high potential above ground; thus a lightning current of 10 kA and a tower footing resistance of 10 Ω will raise the voltage to 100 kV. This voltage, in addition to the service voltage, may cause a flashover between a conductor and tower causing a line-to-ground fault. Figure 4.6b shows the equivalent "ground" network in which R represents the stand-alone resistance to ground of each tower (this resistance may vary slightly between towers but such variations are small and may be neglected) and r is the resistance between two successive towers. These resistances form a ladder network.[2]

If we assume that the line is infinitely long in both directions, then the input resistance of such a network between the points a and b (or c and d) can be shown to be

$$R_{ab} = R_{cd'} = R\tanh\sqrt{\frac{r}{R}} \tag{4.9}$$

and since always $R >> r$ we have that

[2] A.A. Zaky and T.T. El-Sonni, Simplified method for determining the resistance to ground of power transmission towers, *YJES*, Vol.6, pp 1-5, 2013.

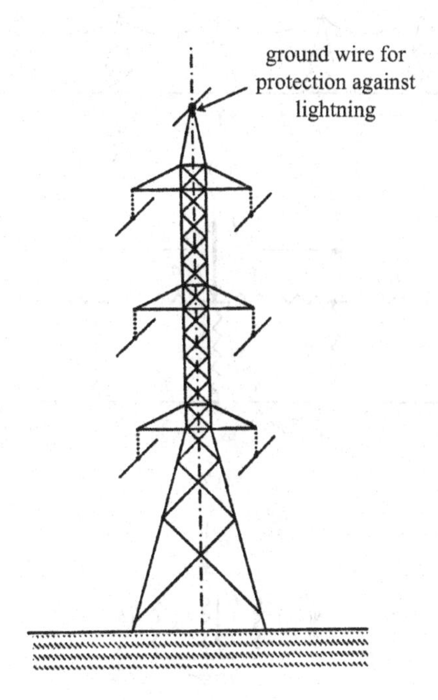

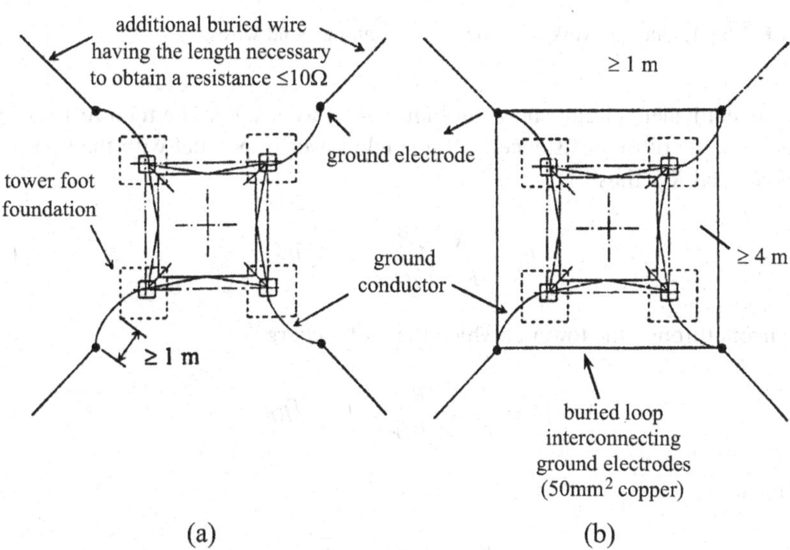

FIGURE 4.5 Grounding of a power transmission line tower.

$$\tanh\sqrt{\frac{r}{R}} = \sqrt{\frac{r}{R}}$$

$$R_{ab} = \sqrt{Rr}$$

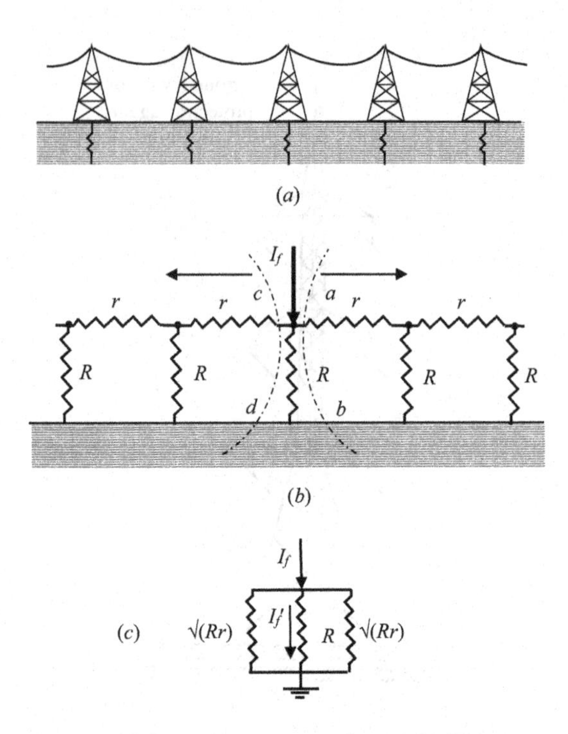

FIGURE 4.6 Ladder network formed by towers and ground wires.

From the equivalent circuit shown in Figure 4.6c we see that the total resistance to ground consists of the resistance R of a single tower in parallel with the two resistances R_{ab} and R_{cd}; thus

$$R_g = \frac{R\frac{1}{2}\sqrt{Rr}}{R+\frac{1}{2}\sqrt{Rr}} \cong \tfrac{1}{2}\sqrt{Rr} \tag{4.10}$$

The current through the tower at which the fault occurs is

$$I_f' = \frac{I_f\frac{1}{2}\sqrt{Rr}}{R+\frac{1}{2}\sqrt{Rr}} \cong I_f\tfrac{1}{2}\sqrt{Rr} \tag{4.11}$$

and hence

$$\frac{R_g}{R} = \frac{I_f'}{I_f} = \tfrac{1}{2}\sqrt{\frac{r}{R}} \tag{4.12}$$

This result indicates that the presence of the ground wire reduces the current through the tower and its resistance to ground by the same ratio. As a numerical example assume that a long power line has a stranded steel ground wire of 10 mm diameter whose resistance is 6.7 Ω/km, that the distance between successive towers is 180 m, and that the footing resistance of each tower is 10 Ω. From Eq.(4.10) we find that

$$R_g = \tfrac{1}{2}\sqrt{10 \times 1.2} = 1.73\,\Omega$$

This shows that the presence of the ground wire leads to a large decrease in the resistance to ground at each tower; this means that the step and touch potentials are greatly reduced as is the danger of flashover. If the towers are close to the ends of the line (less than 2 km) the resistance to ground is

$$R_g = \sqrt{Rr}$$

which is double the resistance of the more distant towers.

5 Grounding Systems

5.1 MAGNITUDE OF RESISTANCE TO GROUND

The ideal value of the resistance to ground is zero. Although in practice this can never be attained, it is possible to obtain values of less than 1 Ω. However, in many cases such low values are not necessary. In general the value of the desirable resistance is inversely proportional to the magnitude of the short-circuit current flowing to earth, i.e., the higher the current the smaller the resistance. Since the reactance of cables is lower than that of overhead lines, short-circuit currents are higher in cables than in overhead lines so that the resistance to ground of installations fed by cables must be lower than that of installations fed by overhead lines.

For generating stations and large substations the resistance to ground should not exceed 1 Ω. For small substations and industrial installations the resistance should be usually less than 5 Ω.

It should be mentioned that the standard specifications of many countries specify the value of the maximum permissible resistance to ground. For example in the USA the National Electric Code (NFPA 70: National Electrical Code) recommends that the resistance should not exceed 25 Ω whereas the German Standard (VDE 0100) specifies 5Ω as the maximum value.

5.2 ELEMENTS OF THE GROUNDING SYSTEM

In general any grounding system consists of the following elements (Figures 5.1 and 5.2):

1. An area of land of suitable earth resistivity and in which earth electrodes can be readily driven or buried.
2. The earthing electrodes.
3. Joints.
4. The grounding conductors which are the conductors used for connecting the electrodes together and connecting the electrodes either to the ground connection provisions or directly to the equipment and structures which are to be grounded.
5. Ground connection provisions. In the case of small installations it is possible to use grounding conductors to connect the ground electrode directly to the equipment to be grounded. However in the case of large installations it is preferable to have special provisions consisting of ground bus-bars and suitable terminals for grounding conductors and equipment grounding connections. These provisions should be mounted on walls or foundations at locations which are easily accessible; this is necessary for inspection purposes.

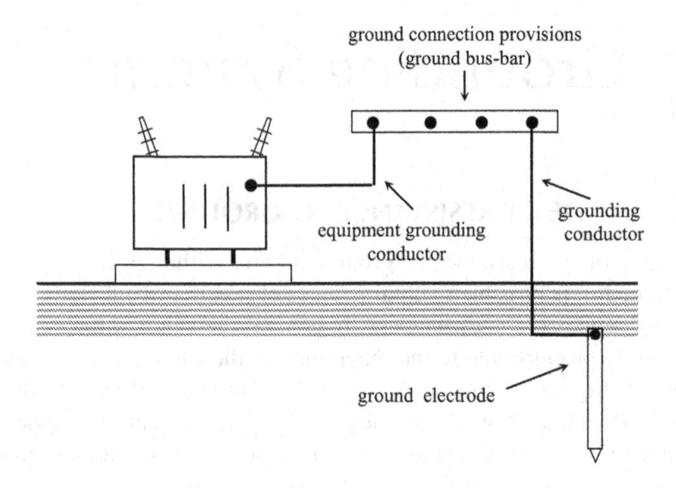

FIGURE 5.1 Elements of a grounding system.

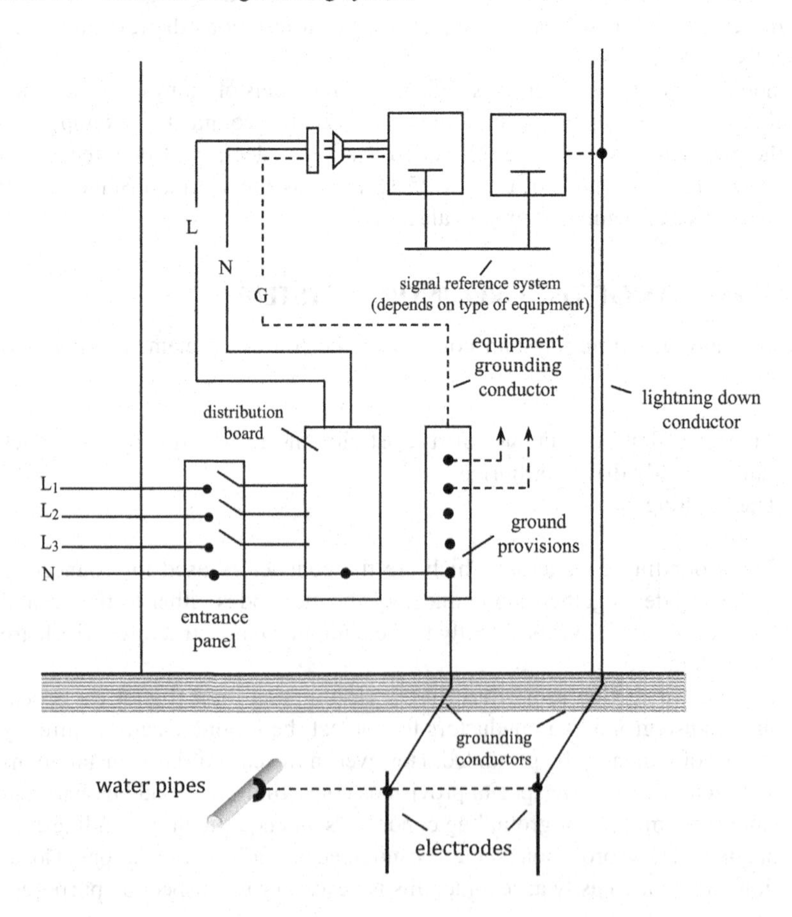

FIGURE 5.2 Elements of a supply system (residential or commercial).

6. Equipment grounding conductors (EGC). These are the connections between the noncurrent carrying parts of the equipment to be grounded and the ground connection provisions referred to above.

It should be mentioned that unlike power supply conductors, grounding and EGC are not protected by inline circuit breakers or fuses. Hence it is necessary to ensure that their cross-sectional areas will safely carry the fault currents for their duration.

5.2.1 The Area of Land

The land must be suitable for placing the grounding electrodes and the resistivity of its soil must be reasonable. If the resistivity is high, the area available is limited, and the driving of rod electrodes to large depths is impossible due to the existence of rocky layers, then it becomes necessary to treat the earth surrounding the electrodes to reduce the soil resistivity.

The soil resistivity depends to a large extent on its moisture content, the amount of salts present capable of forming ions and on its porosity. There are two methods to reduce the resistivity of soils:

- Use of chemical salts.
- Increase the soil's ability to retain water.

5.2.1.1 Use of Chemical Salts

The salts commonly used for improving soil resistivity are, in order of preference,

- Magnesium sulfate ($MgSO_4$)
- Copper sulfate ($CuSO_4$)
- Calcium chloride ($CaCl_2$)
- Sodium chloride—common salt ($NaCl$)
- Potassium nitrate (KNO_3)

Magnesium sulfate is the most commonly used salt because it is cheap, has a high conductivity, and a low corrosiveness. Common salt is also cheap and can be used if there is no danger of corrosion. Potassium nitrate and calcium chloride pose serious health and safety risks and are condemned as hazardous substances by OSHA. Copper sulfate is extremely corrosive to steel, iron, and galvanized pipes, and is considered environmentally unfriendly.

Two common methods used to add chemicals for the improvement of soil resistivity are shown diagrammatically in Figures 5.3 and 5.4. In Figure 5.3 the salt is placed in a trench surrounding the electrode after which the trench is covered with a layer of soil. As an alternative to a trench, a tile pipe of about 20 cm diameter and 60 cm length is buried in the ground surrounding the electrode as shown in Figure 5.4. One half of the pipe is then filled with magnesium sulfate and water is added to fill the pipe. The pipe is provided with removable cover and should be checked periodically and refilled if necessary.

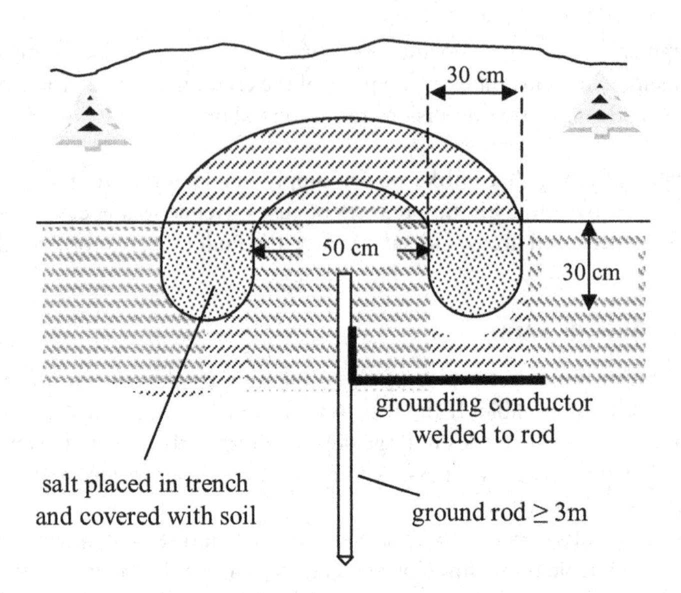

FIGURE 5.3 Salt trench surrounding a ground rod.

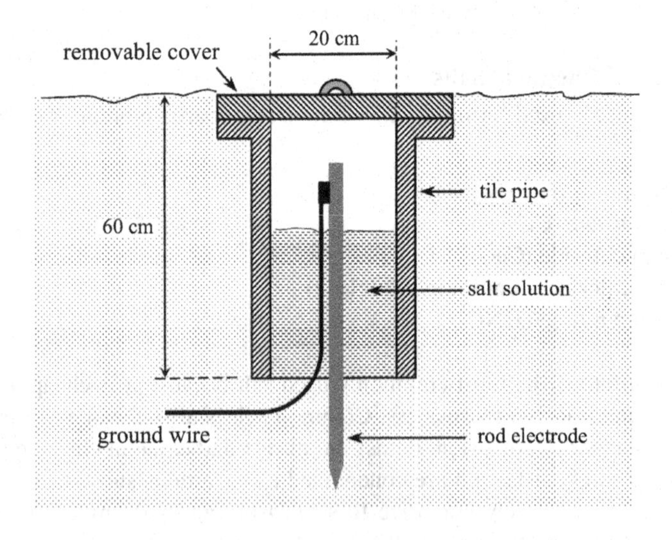

FIGURE 5.4 Treatment with salt solution using a tile pipe.

A more recent method of using chemicals to improve soil resistivity is by using a so-called chemical ground rod electrode (Figure 5.5). This consists of a hollow copper tube of about 6 cm (2.5″) diameter with an exothermically welded pigtail for connection to the grounding conductor. The tube is filled with an electrolytic salt mixture (supplied by the manufacturer); small holes distributed along the tube length allow the hygroscopic salts to absorb water from the surrounding soil and

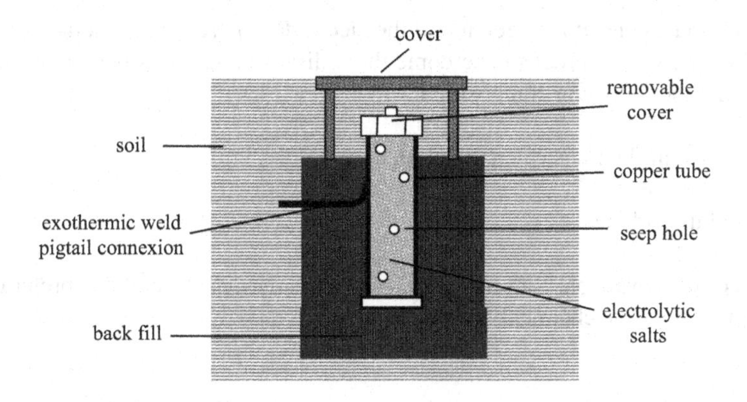

FIGURE 5.5 Chemical grounding rod.

atmosphere. This causes the salts to dissolve and the resulting electrolytic solution seeps out of the hole at the bottom of the tube into the surrounding soil. To further reduce the resistivity the electrode is surrounded by a backfill marketed commercially as a ground enhancement material (GEM) or ground augmentation fill (GAF); these are noncorrosive carbon-based materials that improve grounding effectiveness. It is claimed that a resistivity as low as 0.12Ω·m can be achieved with such installations. Both vertical and horizontal (L-shaped) rods are commercially available; the latter are used at sites where the ground is rocky or for internal installations.

It is apparent that the use of chemical salts for improving the soil resistivity is a temporary measure, since rain and natural drainage will gradually leach away the salt which must be replaced periodically. The frequency of replenishment obviously depends on the amount of rainfall and on the porosity of the soil; on the average it is about every 2 years with the intervals becoming longer as the soil becomes more and more conductive. If follow-up and maintenance is slack, as it is in certain countries, it is advisable not to use this method however economical it might be.

5.2.1.2 Increasing the Soil's Ability to Retain Water

Heavy rainfall and frequent watering lead eventually to the depletion of salts from the soil and a sharp rise in its resistivity. To improve the soil's capability of retaining water, bentonite is added. Bentonite is a kind of soft porous volcanic clay which has the property of absorbing water from its surroundings and retaining it for a considerable time. When placed around earth electrodes, as well as around the conductors joining them together, it reduces the soil resistivity and hence the electrode resistance to ground.

Bentonite is available as a dry powder which is placed directly around the electrodes and the associated conductors. It is then saturated with water and covered with about 30 cm of ordinary soil. One of the disadvantages of bentonite is that when it is saturated with water its volume increases several times, so that care must be exercised when using it in paved areas or near buildings where freedom of expansion is limited; the resulting stresses may lead to dangerous cracks. The other disadvantage is that when bentonite dries its volume shrinks considerably leading to a complete loss of intimate contact between it and the electrodes, thereby prejudicing the entire grounding system. In order to overcome these disadvantages it is best to use a mixture having the following composition:

75% gypsum (hydrous calcium sulfate)
20% bentonite
5% sodium sulfate

This mixture is available commercially, its price is reasonable, and it is preferable to use it rather than use chemical salting.

5.2.2 Ground Electrodes

5.2.2.1 Driven Rods

Since the metal used for the ground electrode does not affect its resistance to ground, the choice of metal will depend entirely on its resistance to corrosion in the type of soil in which it will be buried. Long practical experience as well as laboratory tests have shown that copper is the best metal as far as resistance to corrosion is concerned. The driven rod is the most common type of electrode used for grounding and it is made either of solid copper or of copper-clad steel. The latter is made by molecularly bonding 99.9% pure copper onto a low carbon steel core and is commercially known as a *copperweld* or *copperbond* rod. The steel provides the required mechanical strength which enables the rod to be driven by power hammers to great depths without damage, while the copper cladding protects the rod from electrolytic corrosion and allows the use of a copper/copper junction between it and the grounding conductor. Stainless steel rods are used in applications where soil conditions are very aggressive,such as soils with high salt content, and where galvanic corrosion (see Section 5.3) can take place between dissimilar metals buried in close proximity. Galvanized steel rods are also sometimes used if corrosion considerations permit their use.

Since it is true that no part of an electrical installation is more neglected or forgotten than the grounding system, this part must be designed with great care using only the best materials proven to give the most reliable service for the longest period of time. Copper is such a material and should therefore always be the first choice for earth connections and for electrodes (solid copper or copper-clad).

As the resistance to ground of a rod electrode depends essentially on its length, it has been agreed that the rod length should be equal to or be a multiple of a standard length. This standard length is 8 ft (2.4 m) in the British system and 3 m in the metric system. To drive an electrode to large depths two or more standard lengths are connected together by means of a special coupling. Since the diameter of the rod has no significant influence on its resistance to ground (doubling the diameter will lower the resistance value by only 9.5%) its value is determined by the mechanical rigidity required for driving it into the ground. In general the rod diameter should not be less than 12.5 mm (½″); the size most commonly used is 16 mm ($^5/_8$″).

The most common method of driving the rods into the ground is by using a mechanical hammer especially if the number of rods is large or the soil is not sandy or the length of the rod is more than 3 m. Otherwise a hand mallet can be used to drive the rods.

In most cases it will be found that more than one rod will be needed to obtain the desired resistance to ground. In such cases the distance between the electrodes should not be less than their driven depth. In small transformer substations four or more grounding rods are driven around the perimeter of the installation in the form of a square or rectangle; whenever possible it is preferable to carry this out at the bottom of the site excavations.

In installations with four or more rod electrodes all the rods must be connected together in a closed circuit by means of grounding conductors. The circuit is then connected to the ground connection provisions by means of two or more conductors in parallel.

Figure 5.6 gives typical examples of ground rod connections.

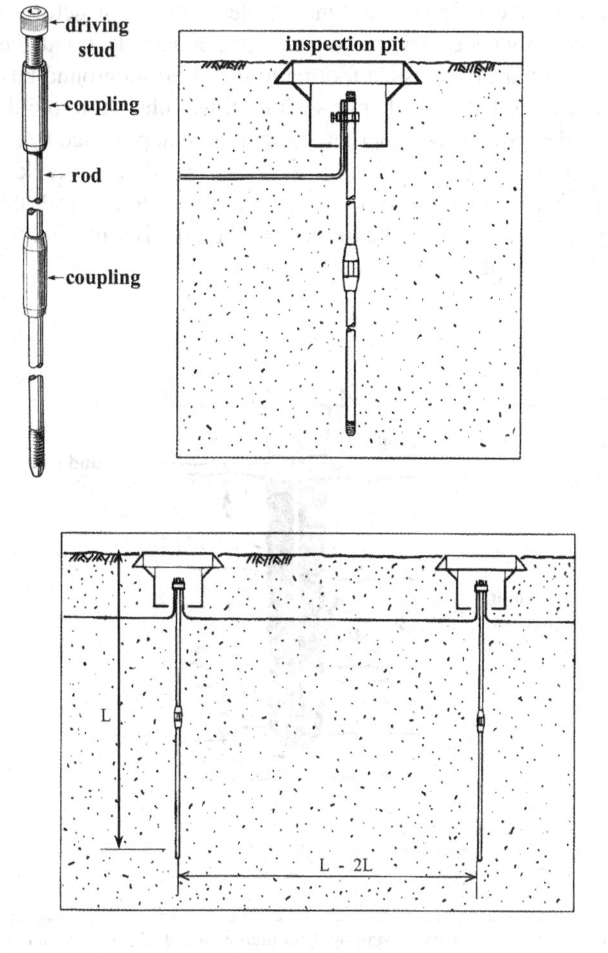

FIGURE 5.6 Constituents of a driven rod and typical installations [Furse].

5.2.2.2 Concrete-Encased Electrodes

The resistivity of concrete below ground level is about 30 Ω·m at 20°C. This is lower than the average value of soil resistivity which is 100 Ω·m. Consequently in earth of average or higher resistivity, the encasement of a rod electrode in concrete (Figure 5.7) gives a lower resistance to ground than a similar electrode buried directly in the ground. This is because the lowering of the resistivity of the material closest to the electrode has the same effect as the chemical treatment of the soil around the electrode (Section 5.2.1).

It should be mentioned here that there are commercial products which when added to cement in place of sand and aggregate produce a conductive concrete whose resistivity is extremely low. One such product is Marconite[1] which is mixed, instead of sand, with ordinary Portland cement in the ratio of 2:1 and then with water to form a fairly dry mix. When used with a copper rod the mix will adhere to the rod and set into a permanent hardened form whose resistivity is as low as 0.1 Ω·m.

The widespread use of steel reinforcing bars in concrete foundations and footings provides a ready-made supply of grounding electrodes at structures utilizing this construction. It is only necessary to bring out an adequate electrical connection from a main reinforcing bar of each such footing to the building ground bus or structural steel (Figure 5.8). The large number of such footings inherent to buildings will provide a net ground resistance considerably lower than that provided by made electrode methods. Concrete-encased steel rods placed in excavations in rock or very rocky soil have been found to be superior to other types of made electrodes. Such types of electrodes constitute the ground electrodes for the majority of steel towers of high-voltage transmission lines.

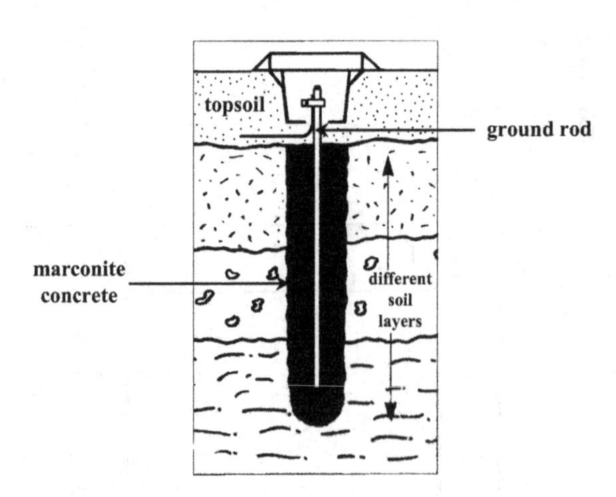

FIGURE 5.7 Ground rod encased in concrete.

[1] Marconite is a registered trade mark of Marconi Communications Ltd., and is manufactured by Carbon Int. Ltd.- Conductive Products Division.

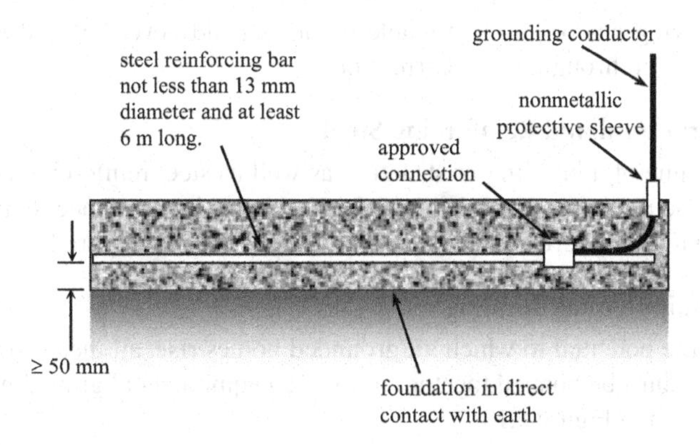

FIGURE 5.8 Concrete-encased bar [NEC 250.52].

5.2.2.3 Water and Gas Pipes

It is possible to utilize water pipes as *auxiliary* earthing electrodes provided that the following conditions are met:

1. The pipes are metallic
2. The pipes are buried in the ground and the buried length exceeds 3 m.
3. There is electrical continuity. If there are junctions or meters they must be provided with suitably sized jumpers as shown in Figure 5.9.
4. The pipes must be connected to the ground electrodes either directly or through the ground bus-bar.

Water pipes should never be used as the sole means of grounding because of the possibility of replacement of some sections by plastic pipes and of non-strict adherence to electrical continuity.

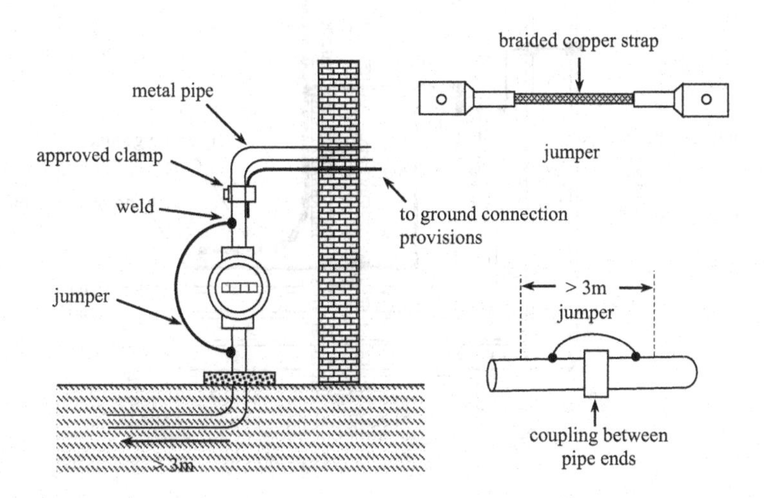

FIGURE 5.9 Jumpers on metal pipes to ensure continuity.

Pipes carrying gas or any flammable liquids should never be used as part of a grounding system through which current flows.

5.2.2.4 Structural and Reinforcing Steel

The metal frame of a building or structure as well as steel reinforcing bars bonded together can be used as a ground electrode. However, the resistance to ground must be measured to make sure that additional electrodes are not required.

5.2.2.5 Equipotential Bonding

To equalize the potential to which all grounded bodies rise, all means of grounding at a given site must be bonded together to form an equipotential grounding electrode system as shown in Figure 5.10.

5.2.3 Connections in Grounding Systems

Joints and connections are critical elements in the grounding circuit through which the short circuit current flows. It is therefore of the utmost importance that they be made with the utmost skill to ensure the integral safety of the grounding system.

There are three methods of connecting the grounding conductors to the ground electrodes, and connecting the EGC to the metal structure of the equipment or to the ground connection provisions. These methods are as follows:

1. Mechanical connection using clamps and bolts.

Both clamp and bolt should be made of the same metal as the electrode and conductor to prevent galvanic corrosion and the connections must be protected against any accidental damage and designed for easy inspection. The most important requirement for

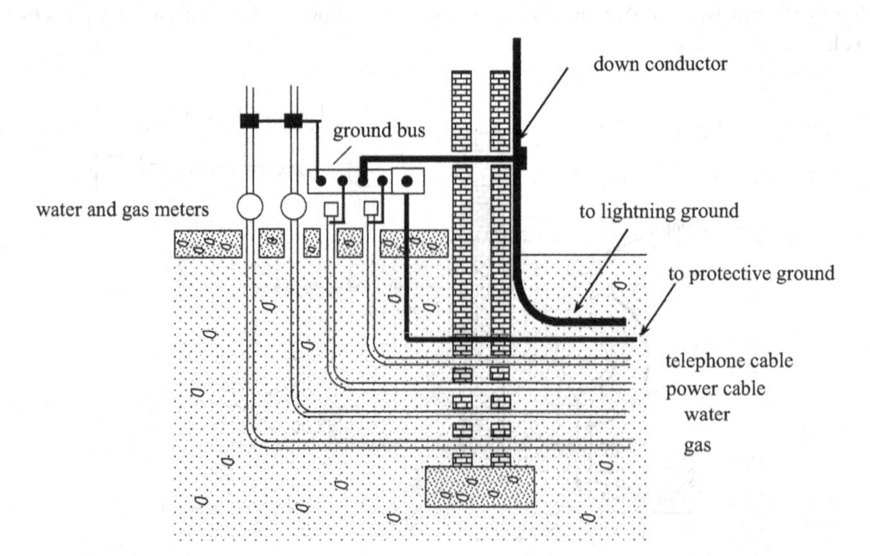

FIGURE 5.10 Equipotential ground system.

clamp connections is the establishment of intimate and permanent contact between the metal surfaces. Therefore, when making a connection the surfaces must be smooth, clean and free from any insulating layer, and the tightening bolts must exert sufficient pressure to maintain good contact in the presence of any stresses or shocks or vibrations associated with the equipment or the surrounding environment. If it is necessary to join two different metals then all necessary precautions (recommended by the manufacturer) must be taken to protect the connection against galvanic corrosion. Figure 5.11 gives some commonly used types of clamps.

2. Welded connection.

This is one of the best methods to make permanent connections and has the advantage that conductors having a smaller cross sectional area can be used (see Section 5.4.2.1). There are several methods used for welding: brazing, silver soldering, and exothermic welding also known as *thermit* welding. Soldering should never be used. All types of welding, except exothermic welding, require experience and high skill.

The best method of welding is the exothermic one and it has the advantage of welding copper to steel or iron. In this method a powder consisting of a mixture of aluminum and copper oxide is placed around the joint inside a graphite mould. When the mixture is ignited by a starting powder, the temperature generated by the exothermic reaction reaches more than 2,000°C and the copper oxide is reduced to molten copper which flows through the mold onto the conductors to be joined. One of the advantages of this method is that it can be used to weld conductors to ground

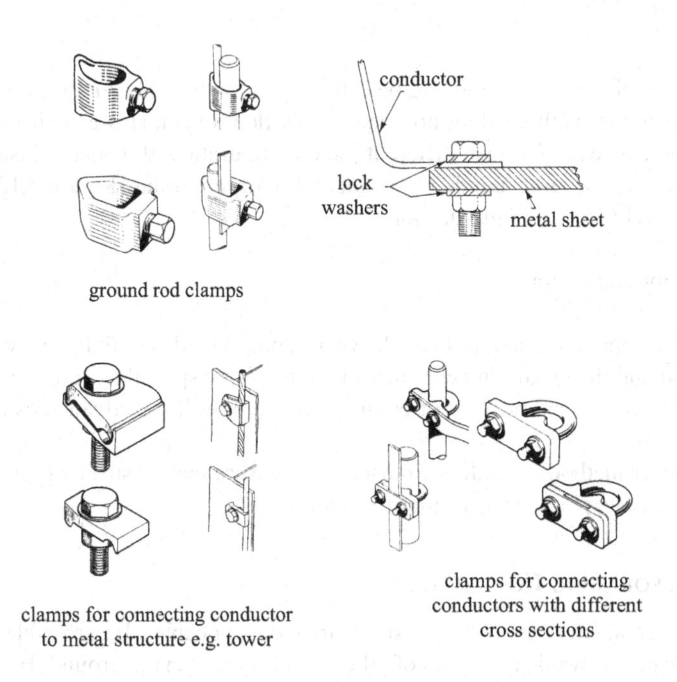

FIGURE 5.11 Some common types of clamps [Furse].

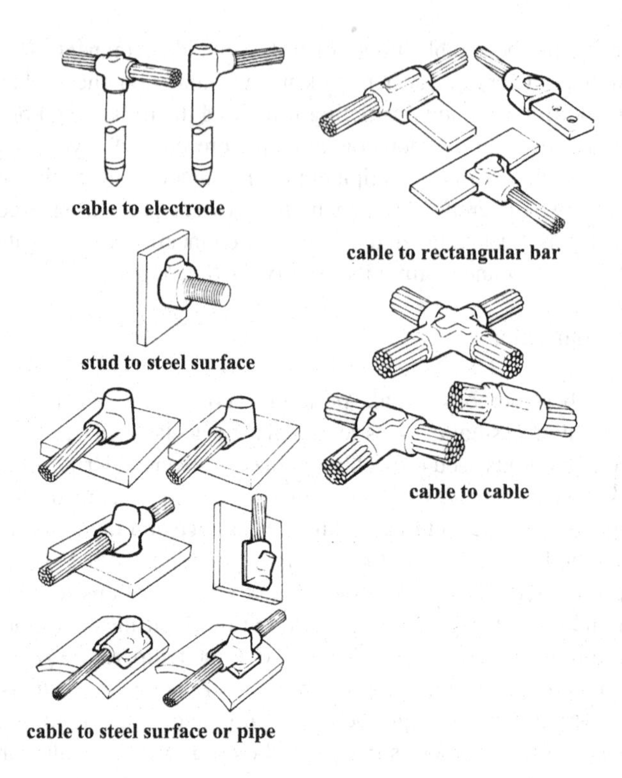

FIGURE 5.12 Welded connections [Furse].

electrodes in places where it would be difficult to use the other methods of welding; another advantage is that it does not require a skilled welder. However, because a different mold is needed for each different joint configuration, this method is economic only if there are a large number of joints which are similar. Figure 5.12 shows a number of welded joint configurations.

3. Compression joint.

A special copper or copper-alloy sleeve is compressed simultaneously onto the ground rod and the grounding conductor by means of a special hydraulic press. This is the most economical method of connection and has all the advantages of thermit welding.

Whichever method is used, one must always ensure that all surfaces are clean and free from grease, paint, or any other insulating film.

5.2.4 Grounding Conductors

Grounding conductors usually consist of buried copper cables. If bare cables are used they may cause galvanic corrosion of other metals buried in the ground. However, for short cable lengths buried near the surface in dry soil of high resistivity, the corrosion aspect may be neglected. Long lengths of copper cable buried in relatively moist soil of low resistivity should have a waterproof cover.

Aluminum, or any other highly anodic metal, should not in any circumstances be used for grounding conductors.

5.2.4.1 Size of Grounding Conductors

Since grounding conductors do not carry current except under short circuit conditions, the choice of the cross-sectional area of the conductor is determined by the magnitude of the fault current and its duration (i.e., the speed with which the protective device disconnects the fault) as well as by mechanical strength if exposed to possible damage. If unprotected then the minimum size to be used is 16 mm^2 in noncorrosive environments and 25 mm^2 in corrosive environments.

It is possible to specify the cross-sectional area of grounding conductors according to either European or US wire sizes. In European specifications areas are given in mm^2 or $in.^2$ (for British specifications) whereas in US specifications areas are given in circular mils. A circular mil is the area of a circle whose diameter is 0.001 in., i.e.,

$$1\,\text{cmil} = 7.854 \times 10^{-7}\,\text{in.}^2 = 5.07 \times 10^{-4}\,\text{mm}^2$$

$$1\,\text{mm}^2 = 1973.5\,\text{cmil}$$

Because a circular mil is such a small unit, the cross-sectional areas of conductors are usually given in MCM units (1 MCM = 1 kcmil = 10^3cmil). For the purpose of comparison, Tables A.1 and A.2 in the Appendix give the cross-sectional areas and diameters of conductors according to the different specifications.

The cross-sectional area of a copper grounding conductor must not be less than the value determined from the following equation (known as the Onderdonk equation) which is based on the assumption that the heating process is adiabatic, i.e., there is no heat lost to the surrounding medium, an assumption which is valid provided that the duration of the short circuit current does not exceed 5 s:

$$I = A'\sqrt{\frac{\log_{10}\left(\dfrac{T_m - T_a}{234 + T_a} + 1\right)}{33t}}$$

$$A' = I\sqrt{t}\,\text{C cmil}$$

$$A = \frac{A'}{1{,}973.5} = \frac{I\sqrt{t}}{\text{k}}\,\text{mm}^2 \tag{5.1}$$

where

A' = conductor area in cmil,
A = conductor area in mm^2,
I = short circuit current (A),

t = current duration (s),

T_a = ambient temperature (assumed 40°C),

T_m = maximum allowable temperature (°C) which is determined by the type of joints,

k, C = constants whose values depends on type of joint. For copper conductors their values are as follows:

C	k	
11.6	170	clamp joints with threaded bolts (T_m = 250°C)
9.11	217	brazed joints (T_m = 450°C)
6.96	284	compression or exothermic soldered joints (T_m = 1,083°C)

The melting point of copper is 1083°C.

The dependence of the conductor cross section on the current duration indicates that the type of protective devices used and their current/time interruption characteristics must be taken into consideration in designing protective earthing (PE) systems.

Example:

Assume that the magnitude of the short circuit current is 20,000 A and that the protective device clears the fault after five cycles, i.e., 0.1 s.

(a) For mechanically clamped joints, $C = 11.6$, $k = 170$,

$$A = \frac{20,000 \times \sqrt{0.1}}{170} = 37.2\,\text{mm}^2$$

$$A' = 20,000 \times \sqrt{0.1} \times 11.6 = 73,365\,\text{cmil}$$

From Tables A1 and A2 we find that
$A_{min} = 50\,\text{mm}^2$
A'_{min} = No. 1 AWG (American Wire Gauge)

(b) For brazed joints, $C = 9.11$, $k = 217$,

$A = 29.14\,\text{mm}^2$
$A' = 57,617$ cmil
From Tables A1 and A2 we find that
$A_{min} = 35\,\text{mm}^2$
A'_{min} = No. 2 AWG

(c) For compressed and exothermic joints, $C = 6.96$, $k = 284$,

$A = 22.7\text{mm}^2$
$A' = 44,019$ cmil

From Tables A1 and A2 we find that

$A_{min} = 25\,mm^2$
$A'_{min} = \text{No.3 AWG}$

It is apparent from this example that compression and exothermic welding give the best type of joint since the conductor cross section for such joints is 50% less than that for mechanical joints and 30% less than that for brazed joints.

5.2.4.2 Protective Sleeve

In some installations it is necessary to protect parts of the grounding conductor path from mechanical damage to which the conductor could be exposed. In these cases the conductor is placed inside a protective sleeve which has sometimes to be a metal one. For these types the sleeve has to be connected to the conductor at both ends as shown in Figure 5.13. The reason for this is that under short-circuit conditions the large current flowing in the conductor produces a strong magnetic field in the sleeve, especially if it is made from a magnetic material such as steel. This field will generate in the sleeve eddy currents which heat it to temperatures that may attain its melting point, causing damage to the conductor. Moreover, the presence of a magnetic sleeve around the conductor causes a large increase in its self-inductance and hence in the impedance of the grounding conductor. To avoid all such effects both ends of the sleeve must be connected to the conductor with wire of the same size as the conductor itself. The majority of metal sleeves consist of steel pipes and since the larger part of the current passes through the pipe (because of the skin effect) its electrical resistance must not be less than that of a similar length of grounding conductor. Table 5.1 shows the approximate equivalent sizes of steel pipe and copper conductor.

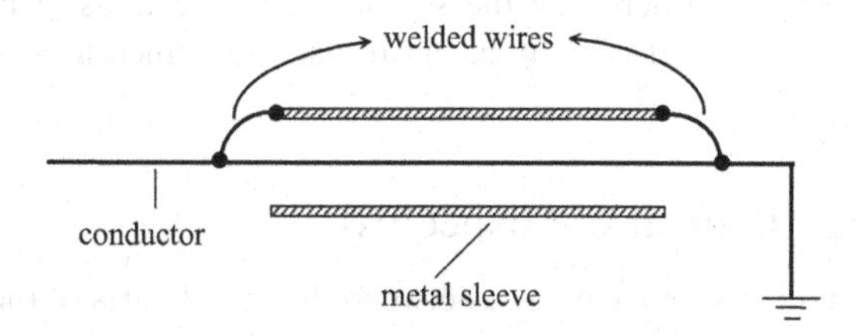

FIGURE 5.13 Protective metal sleeve with jumpers at both ends.

TABLE 5.1
Equivalent Sizes of Steel Pipe and Copper Conductor

Steel pipe (inch)	Copper conductor mm²	Copper conductor AWG
½	25	No. 4
¾	35	No. 2
1	50	No. 1
1 ¼	70	No. 2/0

5.2.5 Ground Connection Provisions

Except for very small installations, it is not feasible to connect the grounding conductors directly from the ground electrodes to the equipment. The normal procedure is to terminate the buried grounding conductor at ground connection provisions. These provisions consist of plates or bus bars located at readily accessible locations in the installation. Ground plates are used in concrete structures such as equipment foundations, vaults, manholes, piers, etc. The plates are placed at a suitable site on the concrete structure such that the flat surface is flush with the concrete surface. Grounding conductors are connected to the plates by means of special clamps to which the conductors are either welded or fixed with bolts. If the installation of ground plates is not practical then copper bus-bars are used. These bars are fixed to the surface of walls at convenient locations for either permanent or temporary ground connections. They must not be fixed to any metal surface and the diameter of the fixing hole on the bar itself must not exceed one quarter of the bar width. The size of the bus bars is determined as for the grounding conductors; for distribution stations at which the short circuit current does not exceed 22 kA, the minimum cross section of the ground bus is $125\,mm^2$ ($5 \times 25\,mm$). Connections to the bus bar may be either mechanical or welded.

In installations where fuses, disconnect switches, cable trays, etc., are mounted on metal beams, these beams may be used as ground connection provisions for such equipment provided that the beams are properly grounded. They may also be used for the temporary grounding of minor equipment not mounted on the beams. However, major pieces of equipment, such as transformers and circuit breakers, have to be grounded through proper ground connection provisions even if they are mounted on grounded beams.

It should be mentioned here that the standard specifications of many countries do not permit the passage of current except in paths confined to a specific electric circuit.

5.2.6 Equipment Grounding Conductors

These are the conductors which extend between the metal parts of the equipment to be grounded and the ground connection provisions referred to above. Several means are used to ground equipment:

1. Bare or covered copper cable
2. Rigid metal conduit or raceway
3. Flexible metal conduit (FMC)
4. Metal cable trays
5. Shield and armor of some cables

The most common means is copper cable.

The cross-sectional area of EGC used in factories and stations is determined by using Eq.(5.1) exactly as for grounding conductors.

In residential and commercial buildings EGC are the protective conductors, which are usually identified by an insulating cover which is green or green with one or more yellow stripes. The method for determining the size of these conductors depends upon which national specification is used. In what follows we shall give three of the most commonly used methods.

5.2.6.1 Conductor Size according to the Adiabatic Equation

The conductor cross section is determined from the following (adiabatic) equation:

$$A = \frac{I\sqrt{t}}{k}\,\text{mm}^2 \tag{5.2}$$

where

A = conductor area in mm^2,
I = short-circuit current (A),
t = current duration (s), which is the disconnection time of the protective device,
k = a factor which depends on the material of the conductor, the insulation and the initial, and final temperatures.

Equation (5.2) is often expressed as

$$A = \frac{aI_n\sqrt{t}}{k}\,\text{mm}^2 \tag{5.2a}$$

where

I_n = the rated current of the protective device and
aI_n = multiples of the rated current.

Equation (5.2) is known as the adiabatic equation because it is based on the assumption that for the duration of the short-circuit current there is no loss (or gain) of heat. This means that the entire ohmic losses generated during the flow of this current are used to heat the conductor without any loss to the surrounding medium. This assumption is valid provided that the duration of the short circuit is less than 5s and since all modern LV circuit breakers and fuses have short circuit tripping times very much shorter than that (a few hundred milliseconds), using the adiabatic equation greatly simplifies calculations. The equation is derived as follows.

For a short-circuit current I with a duration time t the energy dissipated by the ohmic losses is

$$E_r = I^2Rt$$

This energy will be converted to heat raising the conductor temperature by $\Delta\theta$. This thermal energy is given by

$$\begin{aligned} E_{th} &= \text{mass} \times \text{specific heat} \times \text{temperature rise} \\ &= mc\,\Delta\theta \end{aligned}$$

where

m = mass of cable conductor (gm)
c = specific heat of conductor material (J/gm·°C)
$\Delta\theta$ = maximum permissible temperature rise (°C)

Since the process is assumed to be adiabatic the two energies E_r and E_{th} must be equal so that

$$\int_0^t I^2 R dt = mc \int_{\theta_1}^{\theta_2} d\theta \tag{5.3}$$

and assuming that the resistance is constant

$$I^2 Rt = mc\,\Delta\theta$$

where

$$R = \rho_r l / A; \quad m = \rho_d A\, l$$

and

ρ_r = resistivity of conductor (Ω·mm)
ρ_d = density of conductor material (gm/mm³)
A = cross-section area of conductor (mm²)
Substituting these values in the above equation and solving for A

$$A = \frac{I\sqrt{t}}{k}$$

where

$$k = \sqrt{\frac{Q\,\Delta\theta}{\rho_r}} \tag{5.4}$$

and

$Q = c\,\rho_d$ = volumetric heat capacity (J/°C mm³)

For convenience gm and mm units are used to give the cross-sectional area A in mm².

Table 5.2 gives the values of the relevant properties of conductor materials. These values may vary very slightly for copper and aluminum depending on the manufacturing process. For steel, however, these values may vary considerably, resistivity 104–186 μΩ·mm, specific heat 410–500 J/°C. The values given are for high conductivity steel wire used as armor for cables and also serving as PE conductor.

Q may be considered constant since the increase in specific heat with increasing temperature is compensated by an almost equivalent decrease in density.

In order to determine the value of the k factor in Eq.(5.4) it is necessary to determine what value to use for the resistivity ρ_r which is not constant but increases

TABLE 5.2
Conductor Parameters

Conductor Material	Density (gm/mm³)	Specific heat (J/gm°C)	Volumetric capacity Q (J/°C mm³)	Resistivity ρ_{20} (μΩ mm)	Coefficient of Resistivity α_{20} (/°C)	Coefficient of Resistivity α_0 (/°C)
Copper	8.96×10^{-3}	0.3853	3.45×10^{-3}	17.24×10^{-6}	3.93×10^{-3}	4.26×10^{-3}
Aluminum	2.70×10^{-3}	0.921	2.50×10^{-3}	28.26×10^{-6}	4.03×10^{-3}	4.38×10^{-3}
Steel	7.86×10^{-3}	480	3.8×10^{-3}	138×10^{-6}	4.50×10^{-3}	4.95×10^{-3}

with increase in temperature. The variation of resistivity with temperature is approximately given by

$$\rho_\theta = \rho_0 \left[1 + \alpha_0 \theta\right] = \alpha_0 \rho_0 \left(\beta + \theta\right) \tag{5.5}$$

where α_0 is the temperature coefficient of resistivity at 0°C and $\beta = 1/\alpha_0$. Using this value of ρ_θ Eq. (5.3) now becomes

$$\int I^2\,dt = I^2 t = \frac{A^2 Q}{\rho_0} \int_{\theta_i}^{\theta_m} \frac{d\theta}{1 + \alpha_0 \theta}$$

$$= \frac{A^2 Q}{\rho_0 \alpha_0} \ln \frac{1 + \alpha_0 \theta_m}{1 + \alpha_0 \theta_i}$$

$$= A^2 k^2 \tag{5.6}$$

where θ_i is the initial temperature and θ_m is the maximum temperature the insulation can withstand without damage. Substituting for $\alpha_0 \rho_0$ from Eq. (5.5) and assuming that $\theta = 20$°C we obtain finally

$$k = \sqrt{\frac{Q(\beta + 20)}{\rho_{20}} \ln \frac{(\beta + \theta_m)}{(\beta + \theta_i)}} \tag{5.7}$$

This equation is used by a number of standards and the values of k shown in Table 5.3 are derived from it. Table 5.4 gives the values of k for bare conductors also derived from it as given in the standard IEC60364-5-54.

An alternative method (proposed by the author) of determining the factor k from Eq. (5.4) is to assume a resistivity whose value is the average of its values at the temperature limits imposed by cable type and installation conditions as obtained from Eq. (5.5). It is readily shown that this average is given by

$$\rho_{av} = \tfrac{1}{2} \rho_{20} \left[2 + \alpha_{20} \left(\theta_i + \theta_f - 40\right)\right] \tag{5.8}$$

TABLE 5.3

Values of *k* Derived from Eq.(5.6)

Conductor Material	Installation Conditions	Insulation Material and Initial and Final Temperature (°C)					
		PVC		XLPE EPR		Butyl Rubber	
		30°C	160°C	30°C	250°C	30°C	220°C
Cu	Individual insulated conductor or bare conductor in contact with adjacent cable covering	143		176		166	
Al		95		116		110	
Steel		52		64		60	
		70°C	160°C	90°C	250°C	85°C	220°C
Cu	Conductor within a multicore cable	115		143		134	
Al		76		94		89	
		60°C	170°C	80°C	200°C	75°C	220°C
Cu	Sheath or armor used as grounding conductor	141		128		140	
Al		93		85		93	
Steel		51		46		51	

PVC, polyvinylchloride (70°C)—thermosetting; XLPE, cross-linked polyethylene—thermosetting; EPR, ethylene-propylene rubber—thermosetting; and butyl rubber (85°C)—thermosetting.

TABLE 5.4

Values of *k* for Bare Conductors Where There Is No Risk of Damage to Any Neighboring Material by the Temperatures Indicated

Material of Conductor	Conditions	Visible and in Restricted Areas[a]	Normal Conditions	Fire Risk
Copper	Temp. max	500°C	200°C	150°C
	k	228	159	138
Aluminum	Temp. max	300°C	200°C	500°C
	k	125	105	91
Steel	Temp. max	500°C	200°C	150°C
	k	82	58	50

Reproduced from IEC 60364-5-54:2011 with permission.

The initial temperature of the conductor is assumed to be 30°C

[a] The temperatures indicated are valid only where they do not impair the quality of the connections.

Thus

For copper conductors for which,

$$Q = 3.45 \times 10^{-3}\,\mathrm{J/°Cmm^3}, \rho_{20} = 17.24\,\mu\Omega\mathrm{mm}, \alpha_{20} = 3.93 \times 10^{-3}$$

with cross-linked polyethylene (XLPE) insulated individual conductors (30°C–250°C):

$$\rho_{av} = 25\,\mu\Omega\mathrm{mm}, k = 174$$

with polyvinylchloride (PVC) insulated conductor within a cable (70°C–160°C):

$$\rho_{av} = 24.1\,\mu\Omega\mathrm{mm}, k = 114$$

For bare conductor (30°C–200°C):

$$\rho_{av} = 23.7\,\mu\Omega\mathrm{mm}, k = 157$$

For aluminum conductors for which,

$$Q = 2.5 \times 10^{-3}\,\mathrm{J/°Cmm^3}, \rho_{20} = 28.26\,\mu\Omega\mathrm{mm}, \alpha_{20} = 4.03 \times 10^{-3}$$

with XLPE-insulated individual conductors (30°C–250°C):

$$\rho_{av} = 39.05\,\mu\Omega\mathrm{mm}, k = 115$$

with PVC-insulated conductor within a cable (70°C–160°C):

$$\rho_{av} = 39.05\,\mu\Omega\mathrm{mm}, k = 76$$

These values for k are almost identical to those derived from Eq. (5.6). We may thus rewrite Eq. (5.4) as,

$$k = \sqrt{\frac{Q\Delta\theta}{\rho_{av}}} \tag{5.9}$$

It should be pointed out here that Eq. (5.2) for EGC is similar to Eq. (5.1) for grounding conductors except that the factor k is defined differently. They are both based on the adiabatic assumption which is valid for short-circuit durations less than 5 s.

The duration of the ground fault current depends on the time taken by the protective device (circuit breaker or fuse) to interrupt the current. This duration is primarily determined by the tripping characteristics of the breakers and the time-current characteristics of the fuses used and which are normally supplied by the manufacturer.

Figures 5.14 and 5.15 show typical time-current characteristics for circuit-breakers and fuses.

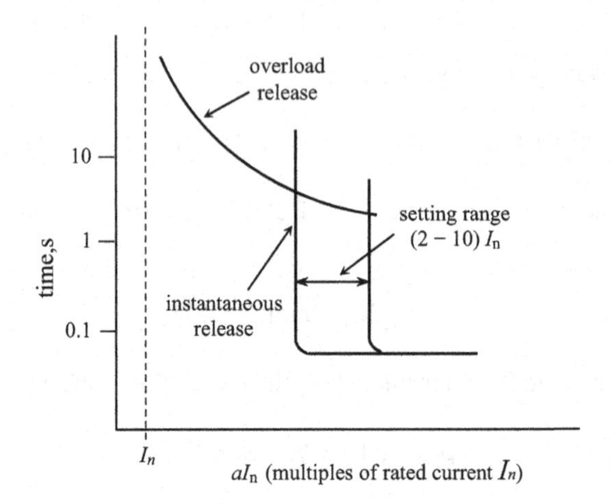

FIGURE 5.14 Tripping characteristics of circuit breakers.

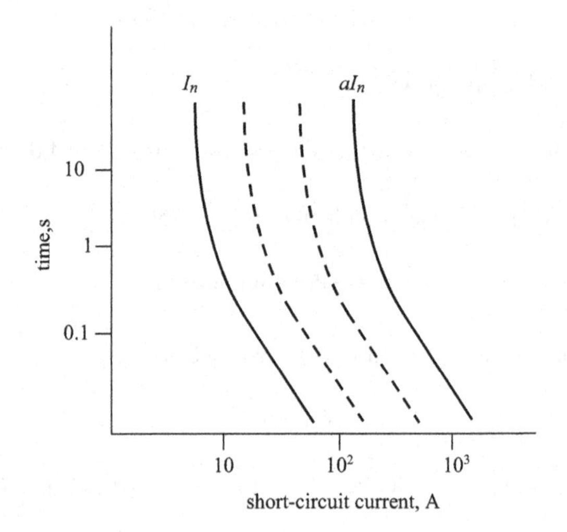

FIGURE 5.15 Time-current characteristics of fuses.

The majority of modern circuit breakers and fuses have fault clearing times around 100ms thus justifying the adiabatic assumption on which Eqs. (5.2) and (5.4) are based. During this period the DC component in the initially asymmetric short-circuit current would have decayed so that the interrupted current may be considered symmetrical.

Short-circuit currents may range from 0.5 to 30 kA depending on the system parameters and distance of fault from the supply source. Table 5.5 shows the grounding conductor size per kA fault current for different values of the factor k given in Table 5.4 and for the time range 0.05–0.5s. For nonstandard cross-sectional areas calculated from the adiabatic equation (5.2) the nearest higher standard section is to be chosen.

TABLE 5.5
Effect of Short-Circuit Duration t on Conductor Cross-Sectional Area per kA Short-Circuit Current

	Conductor Area(mm²/kA)				
k	0.05 s	0.1 s	0.2 s	0.3 s	0.5 s
176	1.27	1.80	2.54	3.13	4.02
166	1.35	1.90	2.66	3.30	4.26
143	1.57	2.21	3.13	3.80	4.94
134	1.67	2.36	3.34	4.09	5.28
116	1.93	2.72	3.85	4.72	6.09
110	2.03	2.87	4.06	4.98	6.42
95	2.36	3.26	4.71	5.77	7.44
89	2.52	3.55	5.02	6.16	7.94
76	2.95	4.16	5.88	7.21	9.30
51	4.39	6.20	8.76	10.63	13.86
46	4.86	6.87	9.71	11.91	15.37

If it is not desired to calculate the minimum cross-sectional area of a protective conductor from Eq. (5.2) then both the standards BS 7671 and IEC 60364 differentiate between two cases:

(a) If the PE conductor is of the same material as that of the phase conductors its cross section is determined as follows:

Phase conductor area (mm²)	PE
$A \leq 16$	A
$16 < A \leq 35$	16
$A > 35$	½ A

(b) If the material is different, then the conductor cross section is chosen such that its resistance is not more than that of the cross section resulting from the application of the above table.

5.2.6.2 Conductor Size According to USA Specifications (NFPA-70)

According to the National Electric Code the cross-sectional area of the EGC is determined by the ampere rating of the protective device (fuse or circuit breaker) installed in the circuit feeding the equipment. Table 5.6 gives the relationship between the size of the grounding conductor and the rated current of the protective device. Figure 5.16 shows the sizes of the different conductors according to this table.

When there is more than one circuit in the same conduit a single EGC may be used (Figure 5.17) and its size chosen according to the rated current of the largest of the protective devices associated with the conductors inside the conduit.

TABLE 5.6
Minimum Size of EGC

Rated Current of Protective Device (A)	Conductor Size[a](copper)
15	14 AWG
20	12
30	10
40	10
60	10
100	8
200	6
300	4
400	3
500	2
600	1
800	1/0
1000	2/0
1200	3/0
1600	4/0
2000	250 MCM
2500	350
3000	400
4000	500
5000	700
6000	800

Reprinted with permission from NFPA 70-2017, National Electrical Code, Copyright © 2017, National Fire Protection Association, Quincy, MA. This reprinted material is not the complete and official position of the NFPA on the referenced subject, which is represented only by the standard in its entirety which may be obtained through the NFPA website www.nfpa.org.

[a] Size not required to be larger than the circuit conductors which supply the equipment (See Tables A.1 and A.2 for relation between conductor size and area).

5.2.6.3 Metal Cable Trays as Grounding Conductors

Cable trays are extensively used in wiring systems in preference to conduits because of their safety features and cost savings. Metal trays are either aluminum or galvanized steel or stainless steel and the ladder type cable tray (Figure 5.18) with 9″ (23cm) between rungs is the most widely used as it provides excellent ventilation for heat dissipation and easy exits and entrances for cables. Metal trays can be used as EGC provided that certain conditions are met. These are given as follows:

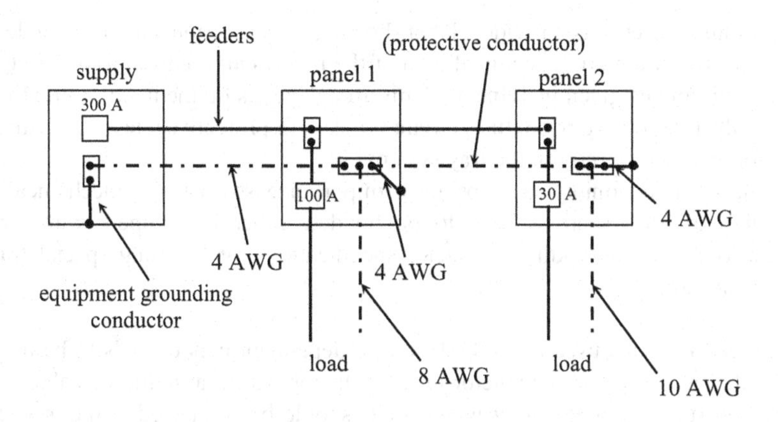

FIGURE 5.16 Size of EGC according to Table 5.6.

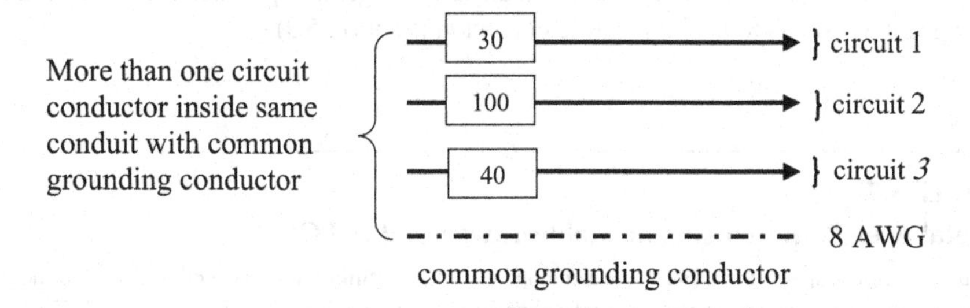

FIGURE 5.17 Several circuit conductors and common protective conductor inside same conduit.

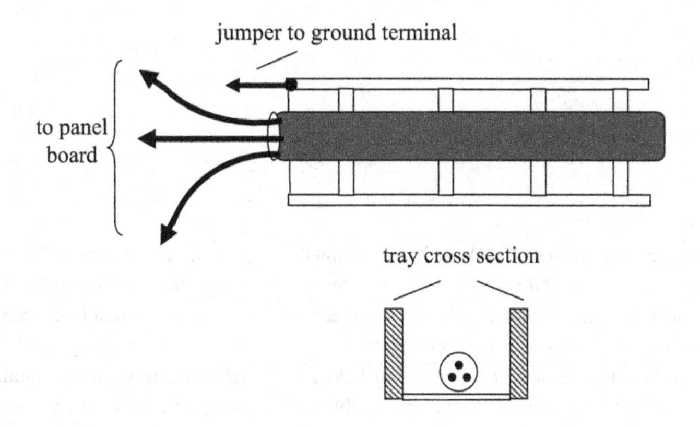

FIGURE 5.18 Cable tray as grounding conductor.

1. Manufacturer must provide a label showing cross-sectional area available.
2. The minimum cross-sectional area of the trays shall conform to the NFPA requirements given in Table 5.7. This area depends on the maximum rating of the protective device (fuse, circuit breaker, or protective relay trip setting) for any cable circuit in the tray system.
3. Electrical continuity is of primary importance so that any mechanically discontinuous sections have to be bonded either by jumpers with size according to grounding conductor specifications, or by using special tray connectors.

If the tray cannot be used as EGC then an independent conductor is to be used as EGC either as a separate conductor or a conductor within a multicore cable. Since the tray must be grounded anyway, the EGC should be connected to the tray at the beginning and at the end and every 50ft (15m); the tray can then act as an auxiliary conductor in parallel with the main EGC.

Connections between metal trays and any copper ground provisions must be made using special strips to avoid galvanic corrosion (Section 5.3).

TABLE 5.7
Metal Area Requirements for Cable Trays Used as EGC

Maximum fuse ampere rating, circuit breaker ampere trip setting, or circuit breaker protective relay ampere trip setting for ground fault protection of any cable circuit in the cable tray system.	Minimum cross-sectional area of metal[a]			
	Steel tray		Aluminum tray	
	in^2	mm^2	in^2	mm^2
60	0.2	1.3	0.2	1.3
100	0.4	2.6	0.2	1.3
200	0.7	4.5	0.2	1.3
400	1.0	6.45	0.4	2.6
600	1.5	9.68[b]	0.4	2.6
1000	–	–	0.6	3.9
1200	–	–	1.0	6.45
1600	–	–	1.5	9.68
2000	–	–	1.5	13[b]

[a] Total cross-sectional area of both side rails of ladder or trough cable trays, or the minimum cross-sectional area of metal in channel cable trays or cable trays of one-piece construction.

[b] Steel cable trays shall not be used as EGC for circuits with ground-fault protection above 600 A. Aluminum shall not be used as EGC for circuits with ground-fault protection above 2000 A.

FIGURE 5.19 Flexible metal conduit (FMC).

5.2.6.4 Flexible Metal Conduits

As the name implies FMC (Figure 5.19) is used wherever flexibility and protection are required, as well as for reducing the number of fittings in an installation or to minimize the transmission of vibrations from equipment such as motors. According to the Institute of Engineering and Technology (IET) Wiring Regulations "flexible or pliable conduit shall not be used as a protective conductor."

However, the use of such conduit as EGC is permitted by the NEC. There are three types of FMC:

1. FMC, which is a conduit of circular cross section made of helically wound, formed, interlaced metal strip.

   Liquid-tight flexible metal conduit—LFMC, which is conduit of circular cross section having an outer waterproof, nonmetallic, sun-light resistant jacket over an inner flexible metal core.
2. Flexible metallic tubing—FMT, which is conduit of circular cross section, flexible, metallic, and liquid tight without a nonmetallic jacket.

The above types can be used as EGC if they meet all the following conditions:

(a) Conditions for all types—FMC, LFMC, and FMT:
   - Fittings used (connectors, couplings, terminations, etc.) are listed for grounding.
   - The total length of flexible piping in the ground circuit does not exceed 1.8 m (6 ft). For greater lengths an EGC must be installed with the circuit conductors.

(b) Conditions for FMC:
   - The circuit conductors contained in the conduit are protected by overcurrent devices rated at 20 A or less.

- If used to connect equipment whose flexibility is necessary after installation, an EGC must be installed.

(c) Conditions for LFMC:

- For ⅜ and ½ in. trade sizes[2] the circuit conductors contained in the conduit are protected by overcurrent devices rated 20 A or less.
- For trade sizes ¾ through 1¼ in. the circuit conductors contained in the conduit are protected by overcurrent devices rated not more than 60 A and the grounding path does not contain any flexible piping of smaller size.
- If used to connect equipment whose flexibility is necessary after installation, an EGC must be installed.

5.2.6.5 Cable Sheath and Armor

The sheath or armor of cables can be used as protective grounding conductor in which case the manufacturer's data must be consulted to determine the sheath and armour electrical specifications (see also Section 5.5.3). It should be mentioned that this also applies to metal raceways used to enclose cables and wires.

5.2.6.6 Jumpers

Jumpers are used to ensure the electrical continuity of the connection to ground of both grounding conductors and EGC as well as to ensure continuity of water pipes used as auxiliary grounding electrodes (see Figure 5.9). Their size is determined in a manner similar to that for the size of grounding conductors or EGC described above.

5.3 METALLIC CORROSION

If two different metals are present in a moist medium or if two different metals are in contact in a moist environment, with the passage of time one of the metals will corrode. The reason for this is the electrochemical process which results in the corrosion of the more anodic or less noble of the two metals. Table 5.8 gives the so-called Galvanic series of some metals and alloys. The Galvanic classification of metals is relative since the measured electrode potentials depend on the electrolytic medium used, but the differences are not large. Table 5.8 also gives the so-called anodic index of the different metals referred to gold as base and for sea water as electrolyte.

A metal is considered to be more anodic than another if it falls above it in the table. For example galvanized steel is more anodic than copper (the voltage difference between them is 0.8 V), but copper is more anodic than gold (voltage difference 0.4 V). If a galvanized steel pipe is buried near a grounding copper electrode, the pipe will corrode but not the electrode.

The rate of corrosion depends on the galvanic potential difference and on the moisture and salt content of the soil or the atmosphere. In coastal areas and polluted

[2] Because the measured size is not the same as the nominal size, 'trade' sizes are used rather than dimensions (e.g., a trade size ½ FMC has an actual inside diameter of 0.635″).

TABLE 5.8
Galvanic Series for Metals and Alloys

Metal	Anodic Index
Anode	
Magnesium and its alloys	1.75
Galvanized steel	1.2
Galvanized iron	1.2
Aluminum	0.9
Cast iron	0.85
Duralumin	0.75
Lead	0.7
Tin	0.65
Tin-Lead solders	0.65
Chromium-plated steel (0.005″)	0.65
Chrome steel 18/2	0.5
Copper and its alloys	0.4
Silver solder	0.35
Nickel-plated steel	0.3
Titanium	0.15
Silver	0.15
Carbon	0.05
Gold	0
Platinum	0
Cathode	

industrial areas the corrosion rate is high. For any two different metals the rate of corrosion of the more anodic of the two is directly proportional to the area of the cathode and inversely proportional to the area of the anode. Hence, to minimize corrosion rate the area of the more anodic metal should not be less than that of the metal cathode (Figure 5.20).

If the ground connections consist of two different metals then the following points must be taken into consideration:

1. The more anodic metal must not be the body of the equipment or of the structure. For example if it is required to connect a galvanized steel structure to copper grounding electrodes, then this must be done by means of a galvanized steel strip which can be easily replaced if it is corroded by galvanic action.
2. Junctions must be above the earth's surface.
3. Junctions must be protected against moisture.
4. Junctions must be located at readily accessible sites for inspection purposes.
5. Junctions must be inspected once a year.

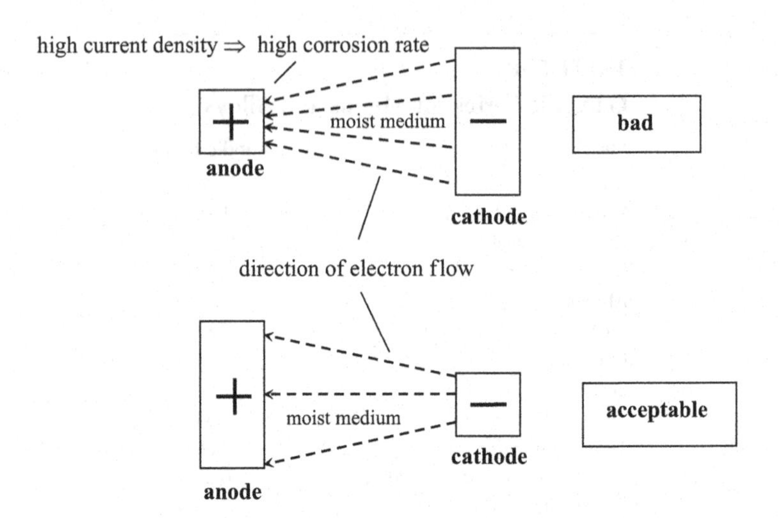

FIGURE 5.20 Effect of anode and cathode areas on corrosion rate.

5.4 GENERAL RULES FOR THE GROUNDING OF BRANCH SUBSTATIONS

Figure 5.21 shows the different grounding systems for a simple network consisting of a branch substation and a distribution substation feeding a low voltage (≤1,000V) network. The different grounding systems are as follows:

A. High-voltage protective grounding system which is the grounding of all metal parts (equipment casings, cable sheaths, etc.) which belong to the high-voltage network.
B. Grounding of the neutral point.
C. Low-voltage protective grounding which is the grounding of all metal parts which belong to the low-voltage network.

The question which now arises is this: should each grounding system be separate from the others, or should a single common grounding system be used? Although

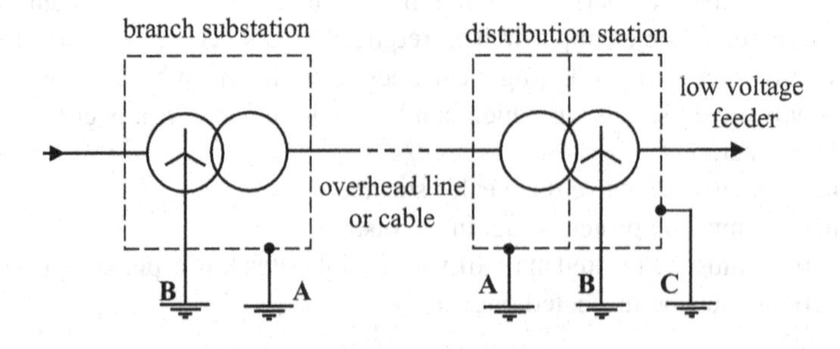

FIGURE 5.21 Different grounding systems at transformer stations.

there is no clear-cut answer to this question, long experience in many parts of the world has led to the adoption of the following recommendations.

I. Transformer substations with no outgoing low-voltage networks

It is recommended to use a common ground for the neutral point and for the protective grounding. The reason for this is as follows.

If a fault occurs inside the station between a conductor and a grounded metal body, then if the two grounding systems are separate (Figure 5.22a) the whole short circuit current will flow into the protective ground. However, if the two ground points do not fall outside each other's resistance area there is bound to be a resistance coupling between them as shown in Figure 5.22b. In order to clarify what the resistances R_1, R_2, and R_m represent let us assume for simplicity that the earth electrodes are represented by equivalent hemispheres of radii a and b, respectively. From Eq. (2.1) the resistance from electrode A to electrode B is given by

$$R_1 = \frac{\rho}{2\pi}\left(\frac{1}{a} - \frac{1}{D}\right)$$

$$= R_a - R_m$$

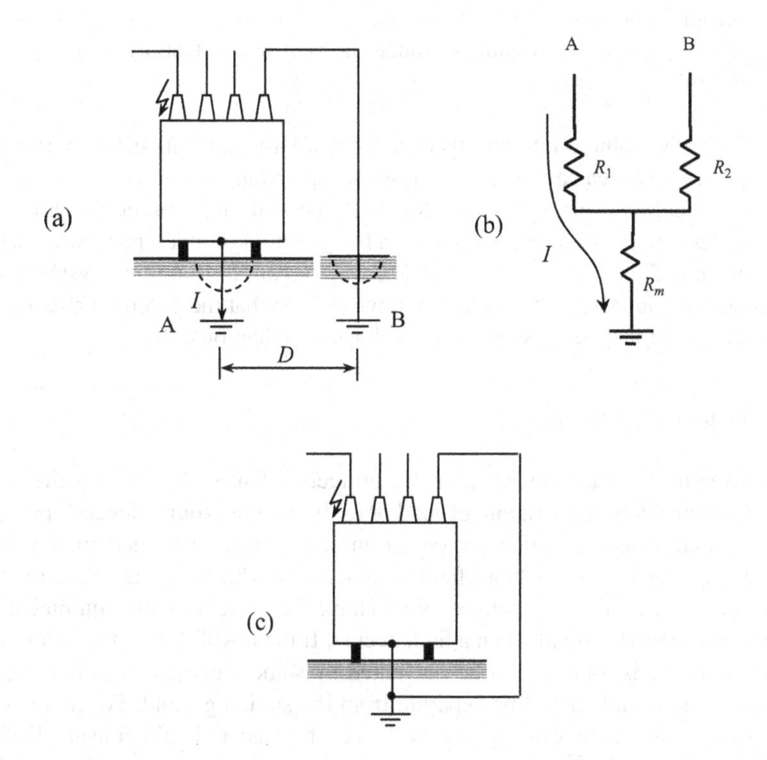

FIGURE 5.22 (a) and (b) separate grounds; (c) common ground.

where $R_a = \rho/2\pi a$ is the absolute resistance of electrode A and $R_m = \rho/2\pi D$ is the resistance from absolute zero (infinity) up to a distance D from A. Similarly the resistance from electrode B to electrode A is given by

$$R_2 = \frac{\rho}{2\pi}\left(\frac{1}{b} - \frac{1}{D}\right)$$

$$= R_b - R_m$$

where $R_b = \rho/2\pi b$ is the absolute resistance of electrode A. Thus,

$$R_a = R_1 + R_m \text{ and } R_b = R_2 + R_m$$

It is evident that when D is large compared to a and b then the mutual resistance R_m is negligibly small.

The potential of the transformer tank T and any metal bodies connected to the same ground will rise to a value given by $I(R_1 + R_m)$, whereas the potential of the neutral point will rise to IR_m. In this case, therefore, it is necessary to either

(a) insulate the neutral conductor from all metal parts such that the insulation withstands the maximum voltage to which the conductor may rise (determined by the magnitude of the short circuit current and that of the resistance to ground), or
(b) ensure that it is not possible to touch the neutral conductor and any metal part at the same time.

It should also be pointed out here that the flow of large currents between two earth points in a limited area can lead to dangerous step voltages.

With a common ground point for both neutral and protective grounding (Figure 5.22c) the only danger is the rise in the potential of metal parts when a short circuit occurs. This danger may be avoided by having a grounding system with a low resistance and fast-acting protective devices such that the potential rise does not exceed the maximum permissible touch voltage (see Section 4.2).

II. The protective fence

The answer to the question whether the protective fence surrounding the station should be connected to a separate ground or to the station ground depends primarily on the relative danger to which persons or animals outside the fence are exposed if they are touching the fence when a fault occurs, and to which persons inside the fence are exposed if they are simultaneously touching the fence and any equipment connected to the station ground when a fault occurs. If the possibility of the latter occurrence does not exist or if it can be prevented by some means, then it is preferable that the fence ground should be separate from the station ground. For more details on the practices of grounding substation fences the reader should consult IEEE Std 80-2000.

III. Transformer substations feeding low-voltage networks with a neutral conductor

The majority of such stations are distribution stations. The decision whether to have separate grounds or a common ground or two common and one separate ground is determined primarily by the requirement that in the event of a fault to earth on the high-voltage side, the potential of the neutral conductor does not rise to a value which may harm people or cause fires. Since distribution substations can have different types of installation arrangements, we shall give in what follows the practice followed for grounding (common or separated) the various types of installations.

(a) If the incoming or outgoing feeders consist of cables with a lead or aluminum sheath covered with a semiconducting layer or a poorly insulating layer and their length in two different directions is 3 km or more (to ensure that the resistance to ground is 1 ohm or less), then it is possible to use a single common point for the substation ground as shown in Figure 5.23.
(b) If the incoming feeder to the station is an overhead line and the outgoing feeders are cables with an insulating (thermoplastic) outer layer, then we can differentiate between the following two cases:
    (i) The outgoing cable feeds a network having a protective multiple grounding (PME) system (see Section 5.5.4). In this case the high-voltage protective grounding must be separated from the low-voltage protective grounding such that each grounding system lies outside the resistance area of the other, i.e., the distance between the two grounds must be at least 20 m (see Section 3.6). The low-voltage protective grounding is connected to the neutral conductor as shown in Figure 5.24.
    (ii) When the substation is enclosed in a steel or reinforced concrete structure, it is usually difficult to separate the low voltage from the high-voltage grounding systems. In such cases a common ground can be used for the two systems as shown in Figures 5.25 and 5.26.

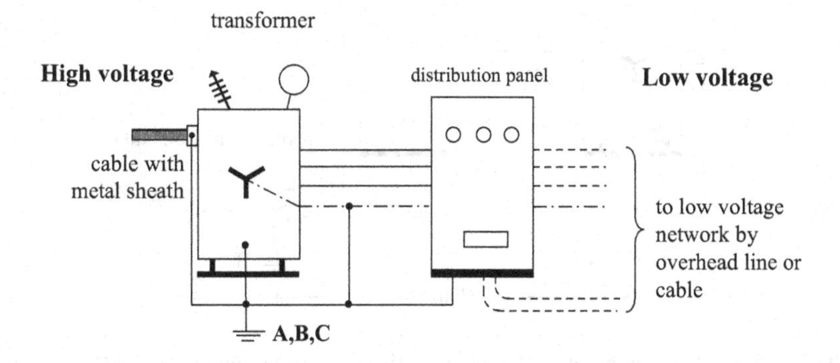

FIGURE 5.23 Input or output feeders cables with metal sheath.

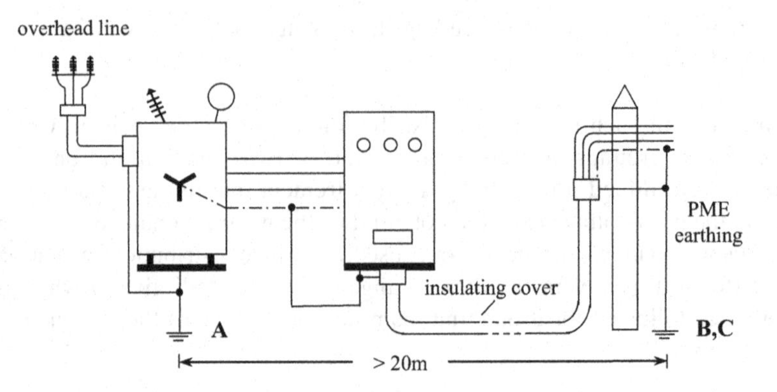

FIGURE 5.24 PME of the distribution system.

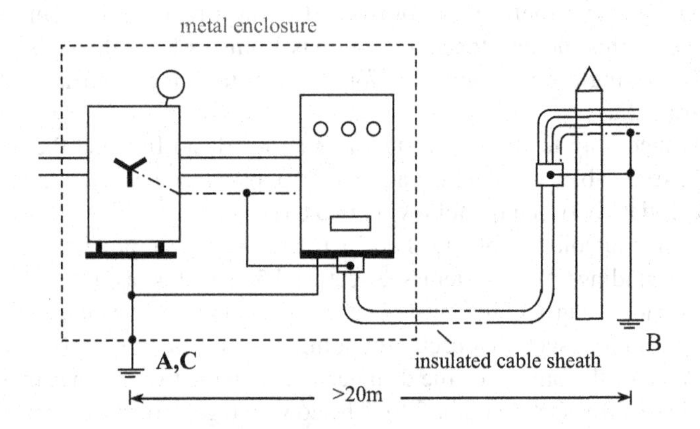

FIGURE 5.25 Common point (A, C) for protective grounding; separate point (B) for neutral grounding.

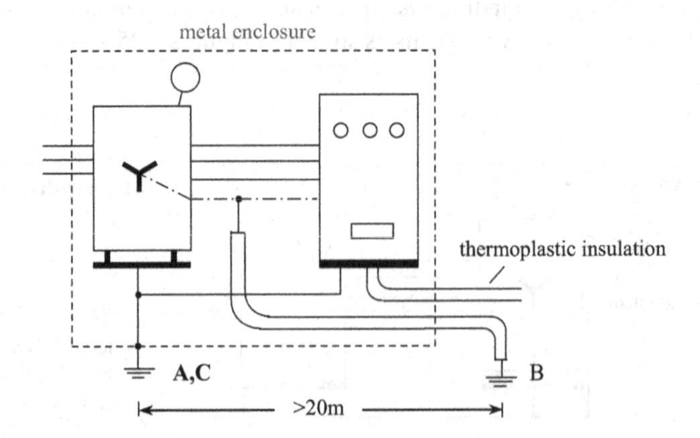

FIGURE 5.26 Common point (A, C) for protective grounding; separate point (B) for neutral grounding.

5.5 PROTECTIVE GROUNDING OF CONSUMER INSTALLATIONS: TYPES OF GROUNDING SYSTEMS

To protect the consumer from electric shock when a fault occurs between a live conductor and a metal body exposed to touch, the following basic requirements must be met:

A. The existence of a closed path in which the short-circuit current can flow. The resistance of this path must be low enough so that the magnitude of the short circuit current is sufficiently high to operate the consumer protective devices (fuses or circuit breakers).
B. The potential of conductive parts exposed to touch, such as equipment enclosures, must not under fault conditions rise to values which are considered dangerous.

According to international standards (IEC 60479-1) the continuous permissible touch voltages are as follows:[3]

For general applications,
≤50 V for alternating current
≤120 V for ripple-free[4] direct current
In agricultural areas and for medical purposes,
≤25 V for alternating current
≤60 V for ripple-free direct current

The admissible limits of touch voltage as a function of time are summarized in Table 5.9.

The types of grounding systems are internationally designated by letters. The significance of these letters is as follows:

First letter: grounding arrangement at the supply source,
T: source has an earth (*terra*) point.
I: source is totally *isolated* or the neutral point is earthed through a high impedance.
Second letter: relationship of exposed metal enclosures and structures to earth,
T: the exposed conductive parts are directly connected to earth irrespective of the grounding arrangement of the supply.
N: direct electrical connection of the exposed conductive parts to the grounding point of the supply, which is usually the *neutral* point.

---

[3] Any circuit in which the potential of a conductor to ground does not exceed these voltages is defined as an extra-low voltage—ELV circuit There are essentially three types of ELV circuits: (a) separated ELV (SELV), which is an *ungrounded* circuit supplied from an isolation transformer, battery, or diesel generator and physically separated from all other circuits; SELV circuitry must be used in swimming pools (see IEC 60364-7-702). (b) Protected ELV (PELV), which is similar to SELV except that it has a protective earth connection for functional reasons. (c) Functional ELV (FELV), any ELV circuit that does not fulfill SELV or PELV requirements.

[4] Ripple-free normally means a rms ripple voltage not more than 10% of the DC voltage.

TABLE 5.9
Maximum Permissible Duration of Touch Voltages

Maximum Breaking Time of Protective Device (s)	Prospective Touch Voltage AC (rms)* (V)	Prospective Touch Voltage DC (V)
∞	<50	<120
5	50	120
1	75	140
0.5	90	160
0.2	110	175
0.1	150	200
0.05	220	250
0.03	280	310

*root mean square

Other letters: these indicate the arrangement of the neutral and protective conductors,

C: neutral and protective conductors *combined* as a common single conductor.

S: neutral and protective functions provided by *separate* conductors.

Types of grounding systems

By system here is meant a single source and all the installations which it supplies.

TN system: In this system the source has one (or more) ground point directly grounded and the exposed conductive parts are connected to that point by a protective conductor. There are three variants of this system:

TN-C: In this system the neutral conductor is also the protective conductor throughout the system.

TN-S: This system has a separate protective conductor throughout the system.

TN-C-S: In this system the neutral conductor is used as protective conductor only in one part of the system.

TT system: The supply has a single earthed point and exposed conductive parts are grounded through an independent grounding system.

IT system: In this system the supply is isolated from earth or connected to earth through a high impedance and all exposed conductive parts are connected to an earth electrode.

Figure 5.27 shows simplified circuits of the above-mentioned systems.

5.5.1 The TN-C System

Figure 5.28 shows several loads fed from a distribution system whose neutral conductor is grounded at the source (transformer). If a fault occurs between a live conductor and a metal enclosure (point 1), the neutral conductor ensures the return path for the

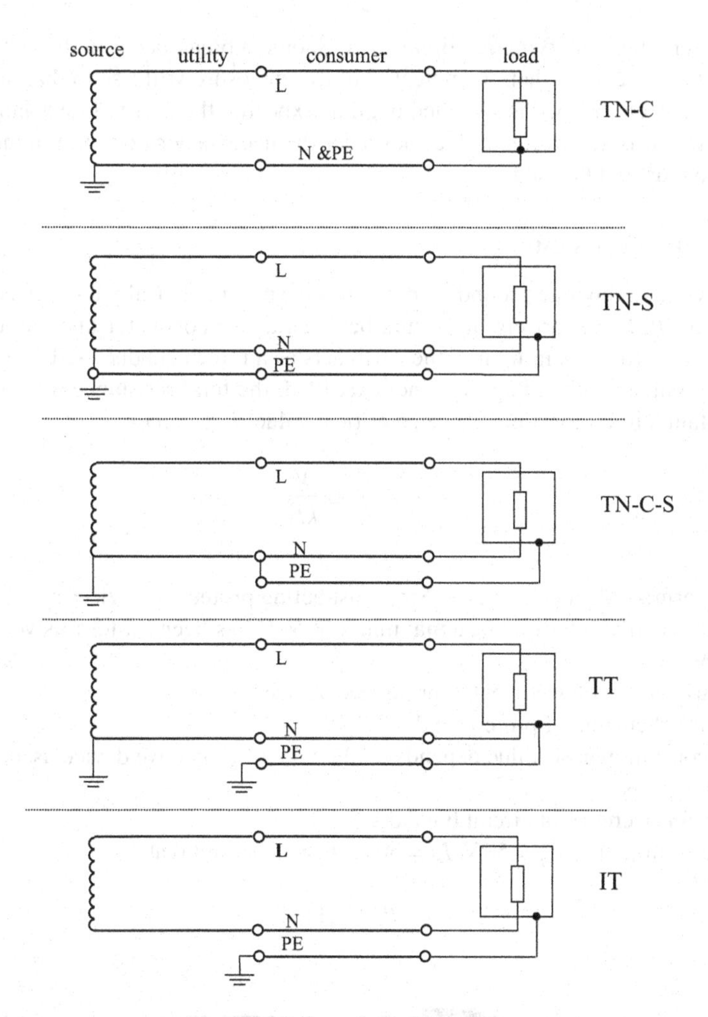

FIGURE 5.27 Different types of grounding systems.

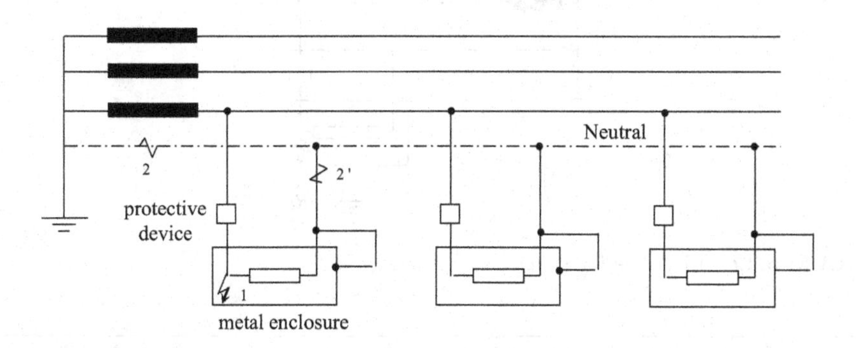

FIGURE 5.28 TN-C grounding system.

fault current thus satisfying requirement A. Should a break occur in the neutral conductor at point 2 or 2′, then the potential of the enclosure will rise to that of the live conductor when the load is switched on, thus exposing the user to grave danger. It is evident that this arrangement does not meet the necessary safety requirements and therefore must not be used.

5.5.2 The TT System

In this system a separate grounding point must be provided locally for each consumer (Figure 5.29). If a short circuit occurs between a live conductor and any conducting enclosure (point 1 in figure), the earth acts as a return conductor. To ensure that the permissible touch voltage V_t is not exceeded, the total resistance $R_c + R_n$ of the ground fault circuit must not exceed a critical value R_t given by

$$R_t \leq \frac{V_t}{I_a} \leq \frac{V_t}{kI_N}$$

where

V_t = permissible touch voltage. Since fast-acting protective devices are not always available for small consumers, a maximum of 50 V has been set for this voltage (see Table 5.9),

I_a = trip current (within 5s) of protective device,

I_N = rated current of protective device,

K = constant whose value depends on the type of protective device used,

= 3 for fuses,

= 1.5 for overcurrent circuit breakers.[5]

If we assume that, $V_t = 50\,\text{V}$, $I_N = 50$ A, $k = 3$, we find that

$$R_t \leq \tfrac{1}{3}\,\Omega$$

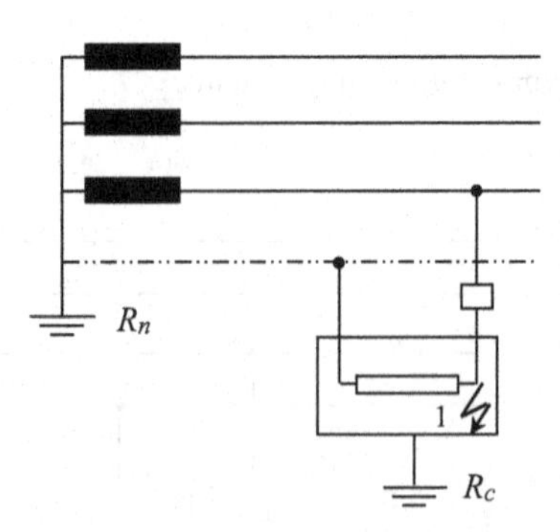

FIGURE 5.29 TT grounding system.

[5] These values for k are typical. The actual value for a particular protective device used can be found from the tripping characteristics of that device as supplied by its manufacturer.

Since it is practically very difficult to achieve such low values for the resistance to ground without excessive costs, this system of grounding is only economical when used in conjunction with current-operated earth leakage circuit breakers (ELCB), also known as residual current circuit breakers. The operating principle of these breakers is as follows (Figure 5.30). The live and neural conductors supplying single-phase loads, and all live and neutral conductors supplying three-phase loads, pass through a toroidal magnetic core onto which is wound a secondary coil connected to the trip circuit. Under normal operating conditions the sum of the currents linking the core is zero and hence no magnetic flux is generated in the core. If there is an earth fault or earth leakage on the load side of the breaker the sum of the currents which link the core is not zero and the net current generates a magnetic flux in the core which in turn induces a voltage in the secondary coil. If the net current is equal to or greater than the rated current of the breaker the tripping circuit disconnects the breaker. The value of the rated leakage current varies between 10 and 500 mA and is chosen according to the type of load and nature of the environment. Regions of high humidity or high pollution require higher ratings than other regions. For the majority of domestic applications the rated current is 30 mA, and for this current, the total resistance to ground R_c may be as high as 1,670 Ω. The operating time of earth leakage breakers varies between 10 and 30 ms.

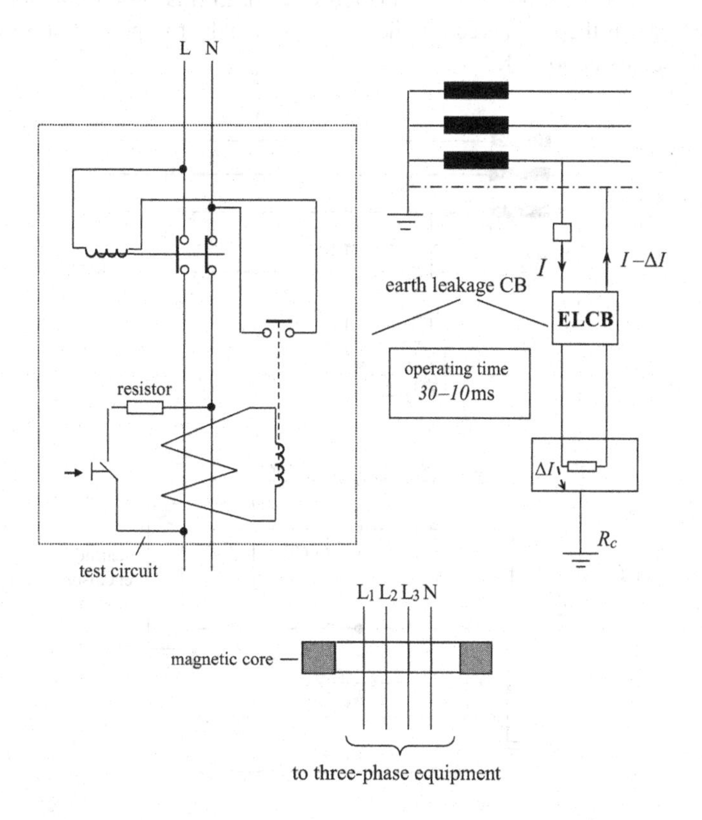

FIGURE 5.30 ELCB.

Because of the high sensitivity of these breakers to any leakage current, they may be used as a protection against the possibility of fire breaking out due to excessive leakage current. The heat generated at the point of fault by the passage of currents smaller than the rated breaker current is insufficient to ignite a fire. These breakers require routine checking and the majority have a self-test circuit incorporated in the design.

ELCBs protect against earth faults but do not provide protection against overcurrent. There are breakers, however, which provide a combination of overcurrent (thermomagnetic) and earth leakage protection (ELCBO).

5.5.3 The TN-S System

In this system five conductors are used instead of four (Figure 5.31). The fifth conductor is the protective conductor and is designated either as G (ground) or as PE.

All exposed metal enclosures and equipment parts which have to be grounded are connected to the protective conductor. This conductor provides the return path for the short-circuit current when an earth fault occurs between a live conductor and the metal enclosure (point 1 in figure); the current returns to the source through this conductor instead of through the neutral as in the TN-C system (Figure 5.28) or through the earth as in the TT system (Figure 5.29). In this system the only danger to a customer is if there is a break in the protective conductor (point 3 or 3′) when an earth fault occurs at point 1.

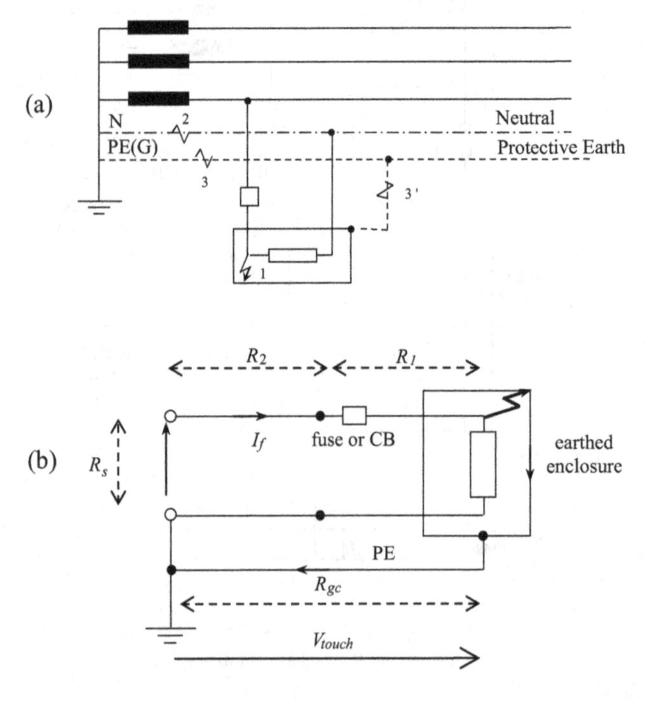

FIGURE 5.31 (a) TN-S system; (b) earth fault with touch voltage $V_{touch} = \text{If}\, R_{gc}$.

Figure 5.31b shows the equivalent circuit for a fault between the live conductor and the metal enclosure. The short circuit current is

$$I_f = \frac{V}{\sqrt{(X_s + X_2)^2 + (R_1 + R_2 + R_{gc})^2}}$$

where

V = source voltage
X_s = inductive reactance of source
X_2 = inductive reactance of supply conductor
R_1 = resistance on conductor from service entrance to fault
R_2 = resistance on conductor from source to service entrance
R_{gc} = resistance of protective conductor from equipment to earthing point.

The reactance of the source can be neglected so can that of cables whose cross section is not greater than 35 mm² (the reactance of a 35 mm² conductor of a three-core cable is 0.092 Ω/km while its resistance is 0.627 Ω/km) so that the fault current is

$$I_f = \frac{V}{R_1 + R_2 + R_{gc}}$$

Now the magnitude of the touch voltage, which is the voltage to which the enclosure will rise under fault conditions, is $V_{touch} = I_f R_{gc}$, and its duration is determined by the time/current characteristics of the protective device.

For miniature circuit breakers with instantaneous tripping (Figure 5.32) the disconnection time does not exceed 0.1s (a tripping time less than 0.1s is considered instantaneous). This time is within that specified by international regulations which

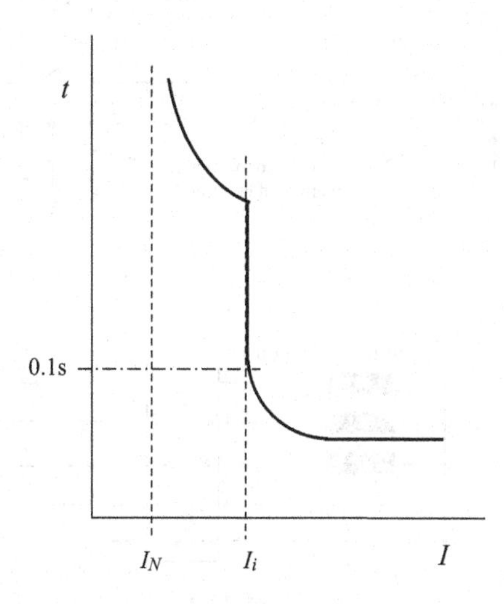

FIGURE 5.32 Time/current characteristic of miniature circuit breaker with instantaneous tripping.

require that for circuits supplying fixed equipment the maximum disconnection time is 5s, whereas for circuits supplying socket outlets the disconnection time shall not exceed 0.4 s. The reason for this is that socket outlets may be used to supply a hand-held equipment requiring that it be gripped continuously in operation (e.g., hand drill) which constitutes a much greater risk to the user than fixed equipment. Now setting

$$I_f = a\, I_N = I_i$$

where

I_N = rated breaker current

I_i = instantaneous tripping current

and for V_{touch} = 50 V, we have that

$$R_{gc} \leq \frac{50}{I_i}$$

Having determined R_{gc} and the fault current I_f, the time t for disconnection of this current is determined from the relevant time/current characteristics of the breaker used. Substitution for I_f, t, and the appropriate value of k in Eq.(5.2) gives the minimum cross-sectional area of the protective conductor and this has to be equal to or less than the size determined by its resistance R_{gc}.

Figure 5.33 shows the neutral (N) and protective (PE) conductors for single-phase and three-phase supply systems when the premises are fed from a general distribution system and when fed from a transformer on the premises. In all cases one must

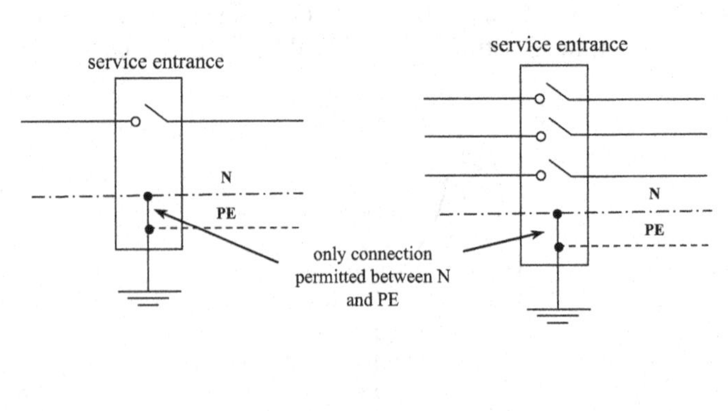

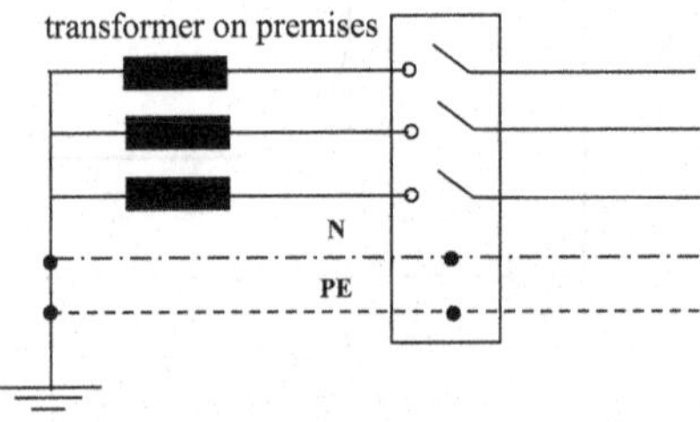

FIGURE 5.33 Connection between neutral and protective conductors.

make sure that there are no other connections between the neutral and protective conductors on the load side of the supply. The reason for this is that should there be another connection, then one part of the load current will return through the protective conductor. This is absolutely forbidden during normal operation because in this case the potential of the metallic parts connected to the protective conductor will rise to a value equal to the voltage drop in that conductor (current × resistance).

It is possible to use the metallic covering (lead sheath and armour) of the cable supplying the installation as the protective conductor. In this case it must be ensured that all junctions are capable of carrying the current flowing in the cable covering. If nonmetallic junction boxes are used the electrical continuity of the covering must be ensured by means of jumpers whose resistance must not exceed that of the covering which has been removed. If the cable is only armored, then in the majority of cases (except if the armor has copper wires) it is difficult to obtain an armor resistance which is sufficiently low to allow the return current to operate the consumer's protective device when an earth fault occurs.

5.5.4 Protective Multiple Earthing

This is a modification of the TN-C system. In this system also the neutral conductor is used as the return conductor for the fault current (Figure 5.34). However, to avoid the drawbacks of the TN-C system the neutral conductor is earthed at the transformer, at the end of the feeder, and at several points along its length (about three points per kilometer) such that the resistance of the neutral conductor between any point and ground does not exceed 10 Ω (or any other value specified by the Authority responsible for the distribution).

It is evident that a break in the system neutral conductor (point 2 in figure) does not constitute a danger to the consumer in the event of a short circuit between a live conductor and a metal enclosure (point 1) since in this case the ground acts as a return conductor for the fault current. However, if a break occurs in the consumer neutral conductor (point 2′) then under fault conditions the metal enclosure's potential will rise to that of the live conductor when the equipment is switched on. To avoid this it is preferable to use a separate protective conductor connected to a

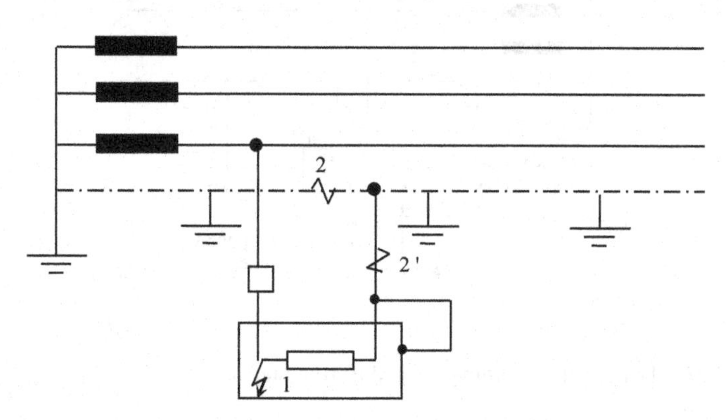

FIGURE 5.34 PME system.

consumer installation earth which is electrically independent of the source earth (Figure 5.35). In this case there is no danger to the consumer except when a double fault occurs: a break at 2′ and a short circuit at 1. However, a break at 2′ in the PE conductor is much less likely to occur than a break at 2′in the neutral connection in Figure 5.34.

It should be stressed that under no circumstances should a direct earthing system be used with any PME system. The following example shows why.

When a fault occurs at point 1 in Figure 5.36, a current of 230/11 = 21 A flows to ground and the potential of the neutral conductor rises to 21 × 10 = 210 V. This potential appears on all metal bodies connected to the neutral conductor.

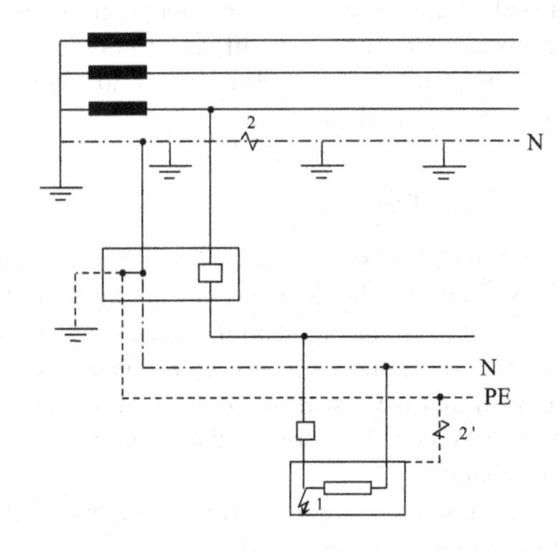

FIGURE 5.35 Use of separate PE conductor in the PME system.

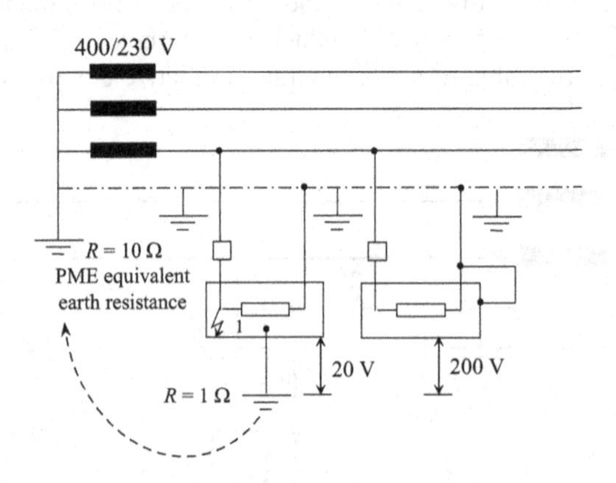

FIGURE 5.36 Danger of combining TT and PME systems.

5.5.5 The IT System

In this system the neutral point at the source is either completely isolated or grounded through a high resistance (impedance) and exposed equipment parts are grounded locally as shown in Figure 5.37. The IT system is only used where continuity of supply is of vital importance such as in some chemical plants with continuous processes and in some medical applications as well as when an extremely low value of first earth fault current is required. In this system the equipment is grounded either through a common protective conductor (Figure 5.37a) or through an individual grounding arrangement (Figure 5.37b). In both cases a first fault between a conductor and ground (earthed enclosure) does not affect the continuity of supply. Should a second fault occur between another conductor and ground the protective device must disconnect the faulted circuit. It is preferable that the protective device is an ELCB backed by fuses and it must be made certain that the touch voltage does not exceed 50 V.

For such systems the majority of international specifications recommend the installation of an earth leakage monitor. This device monitors the condition of the insulation resistance between the isolated system and ground; if this resistance falls

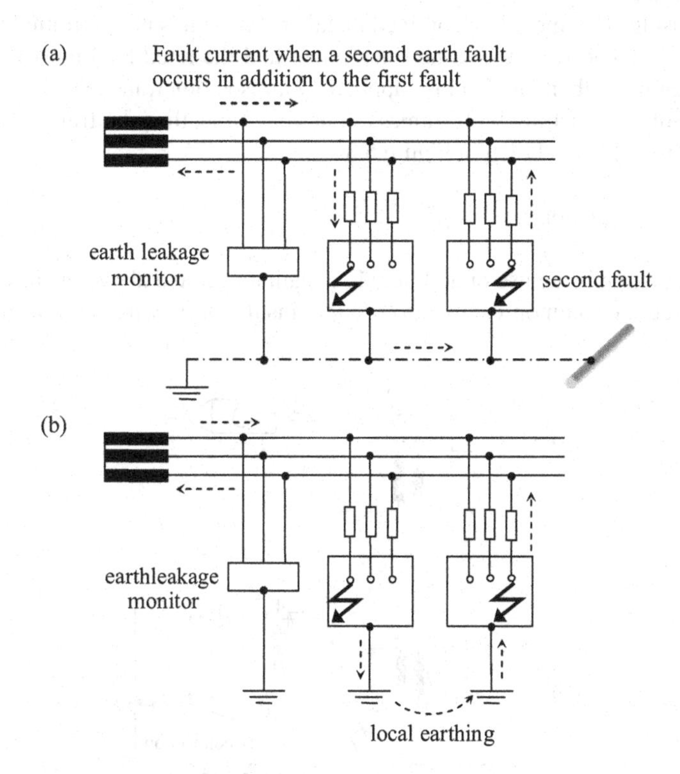

FIGURE 5.37 The IT system: (a) using a common PE conductor; (b) using individual grounding conductor.

below a preset value (between 50 and 200 k·Ω depending on the nature of the installation), the device sets off a sound alarm to warn those responsible of the existence of a fault so that it may be located and removed without having to disconnect the supply.

5.6 PROTECTION BY ISOLATION

(a) Isolation transformers

The load is isolated from the supply source by an isolation transformer. This is a transformer with two completely isolated windings and a 1:1 transformation ratio. These transformers are used for supplying power to portable equipment, especially hand-held tools, to protect users should an earth fault occur between a live conductor and the casing. Figure 5.38a gives an example of the use of an isolation transformer. If the equipment is being used on a metal platform or scaffolding then, in order to prevent the passage of current through the body of the operator in case of a double fault, the equipment casing must be connected to the metal structure as shown in Figure 5.38b.

When using equipment in a confined metal enclosure it is not permitted to supply more than one load from the same transformer and the rated load must not exceed 16 A. When more than one load is supplied by an isolation transformer and there is the possibility of touching both frames at the same time, then the frames have to be connected together as shown in Figure 5.39.

(b) Double insulation

Double insulation is the use of additional insulation over and above the basic insulation required. A common example of double insulation is wire with an insulating

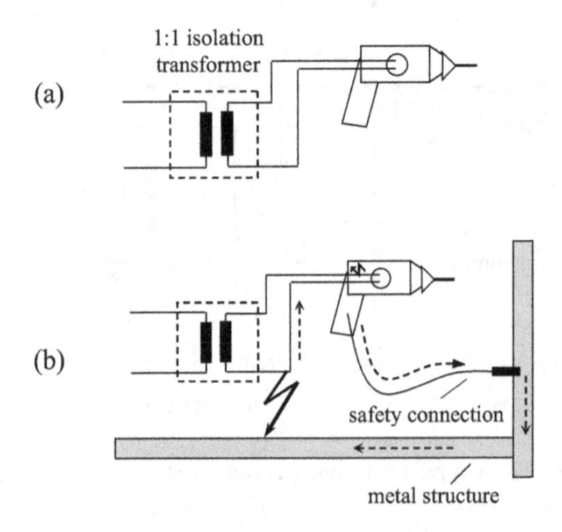

FIGURE 5.38 Use of an isolation transformer as a safety measure.

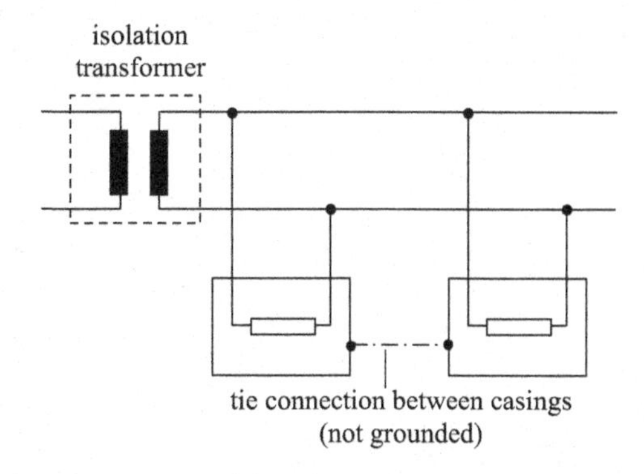

FIGURE 5.39 Isolation transformer supplying more than one load.

cover placed inside a second insulating cover. For electrical machinery and equipment double insulation is accomplished either during the manufacturing or assembling stages or by placing it before use inside an insulated enclosure. Equipment provided with double insulation must conform with the norms and standards applicable to such equipment.

5.7 GROUNDING OF COMPUTERS AND DATA PROCESSING EQUIPMENT

For safety purposes all standards recommend the grounding of all exposed metal casings and enclosures as well as the signal reference plane (SRP) and that there be only a single common ground point. Figure 5.40 gives the grounding elements of a computer or data processing center although it must be emphasized here that the grounding of electronic equipment in particular is coupled with electromagnetic interference (compatibility) and as such is a subject on its own. For additional information on this topic the reader may consult references [17] and [31].

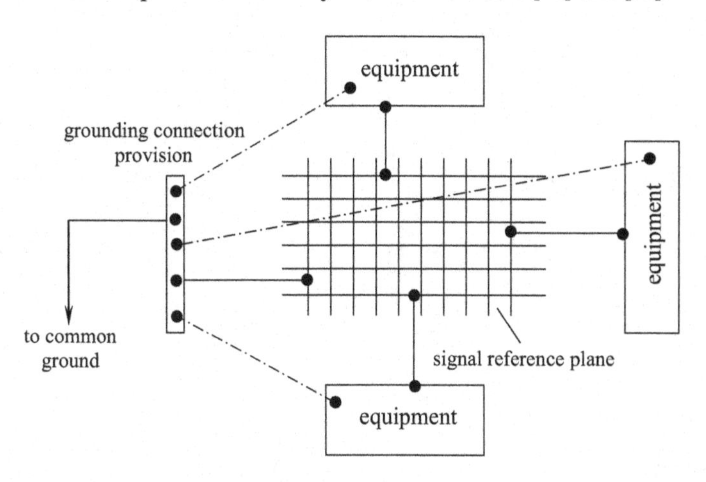

FIGURE 5.40 Grounding elements for a computer center.

6 Substation Grounding Systems

6.1 INTRODUCTION

Substations are an essential part of any electrical power grid and their grounding system must be designed to ensure a safe environment for people who happen to be in and around the substation during a ground fault. This is achieved by limiting the ground potential rise (GPR)—and hence the resulting step and touch voltages—to a value that is below the hazard level for human beings. For large substations the resistance of the grounding system should not exceed 1Ω while for smaller stations it should not exceed 5 Ω.

The basic substation ground system used by most utilities[1] is a grid system consisting of bare copper cables buried horizontally about 30–40 cm in the ground and forming a network of squares or meshes (Figure 6.1). Such a system usually extends over the entire substation area. The spacing of the conductors (mesh size) varies according to the requirements of the installation.

A grid is used instead of the conventional driven rods for the following reasons:

(a) As discussed in Chapter 4 the actual step and touch voltages depend on the magnitude of the fault current and on the resistance to ground of the earthing system, i.e., on the voltage gradient in the station ground. In large substations the fault current is usually very high and it is difficult to obtain a ground resistance which is sufficiently low to ensure that the potential rise will not attain values which would endanger personnel. The potential gradient may be controlled by a properly designed ground grid.

(b) In the grounding of substations it is always necessary to use multiple grounding electrodes as no single electrode would be able to carry the very high short-circuit currents. The conductors interconnecting these electrodes

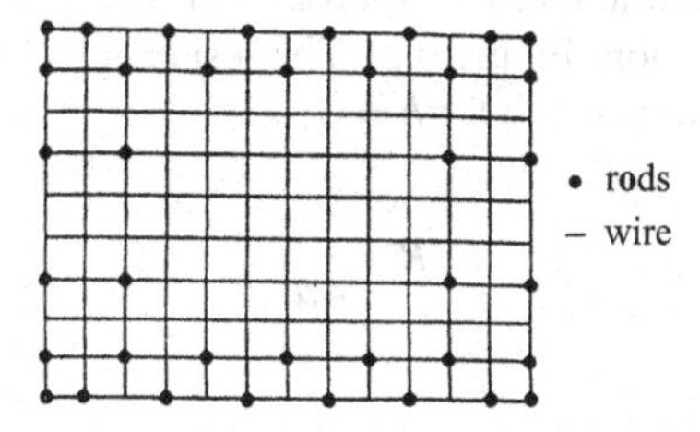

FIGURE 6.1 A typical meshed ground grid.

[1] The outline given in this chapter is based essentially on IEEE Std.80-2000: *Guide for safety in AC substation grounding.*

form a grid or mesh network which in itself constitutes a grounding system. In a soil of reasonably low resistivity it has been found that this network may be so effective as to make the original driven rod electrodes unnecessary.

Connections between the various ground leads and the wire grid as well as connections at the crossovers within the grid are usually clamped, brazed, or welded. Crossovers should be brazed or welded. Ordinary soldered connections are to be avoided because of failure under high fault current or because of galvanic corrosion. Each element of the ground system has to be designed so as to resist fusing and deterioration of electric joints under the worst ground fault conditions using Eq.(5.1).

A typical grid system usually extends over the entire substation area and sometimes beyond the fence which surrounds the building and equipment. Since the fence is usually located on the periphery of the grid area where surface potentials are highest, the grounding of the fence is of major importance especially since the outside of the fence is usually accessible to the general public. It is therefore preferable to extend the area of the ground mat such that the fence lies within it. Readers should consult Ref. [7] for a full discussion of fence grounding.

6.2 DESIGN STEPS

The design engineer will draw up a preliminary design for the grid and then check that it meets the safety requirements. If not, the design shall be modified (repeatedly if necessary) until all safety requirements are satisfied. The design involves the following steps:

1. Determination of soil resistivity
2. Computation of mat resistance to ground
3. Computation of maximum potential rise
4. Computation of touch and step voltages
5. Ensure that safety requirements are satisfied

6.2.1 Ground Mat Resistance

The ground mat of a substation may be approximated by a circular disc whose area is equal to that of the substation (Figure 6.2). The resistance of the disc can be obtained from Eq.(3.5) given in Section 3.4, for $h = 0$,

$$R = \frac{\rho}{4a} \tag{6.1}$$

where

R = resistance of mat to ground (Ω)

ρ = average ground resistivity (Ω·m)

a = radius of disc of same area as that occupied by mat (m)

Assuming that the mat area is A m^2, the radius of the disc is

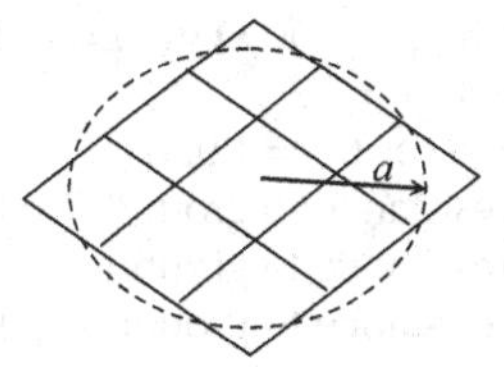

FIGURE 6.2 Representation of a ground mat by an equivalent disc.

$$a = \sqrt{\frac{A}{\pi}}$$

so that

$$R = \frac{\rho}{4}\sqrt{\frac{\pi}{A}} \tag{6.2}$$

This equation is usually modified by adding a second term as follows:

$$R = \frac{\rho}{4}\sqrt{\frac{\pi}{A}} + \frac{\rho}{L} \tag{6.3}$$

where L is the total length of buried conductors in meters. This second term takes into consideration the fact that the ground mat resistance is greater than that of a solid circular disc and that this difference decreases as the length of the conductor increases (for a solid disc $L = \infty$).

Equations (6.2) and (6.3) are approximate and there are more accurate alternative formulas for estimating the ground mat resistance. Sverak's formula[2] which takes into account grid depth is

$$R = \rho\left\{\frac{1}{L} + \frac{1}{\sqrt{20A}}\left(1 + \frac{1}{1 + h\sqrt{20/A}}\right)\right\} \tag{6.4}$$

where h is the depth of the ground mat, usually between 0.25 and 2.5 m.

Example 6.1

Figure 6.3 shows a ground mat whose area is 60 × 60 m.

The ground mat consists of 4 × 4 m meshes and is buried at a depth of 0.3 m. The conductor size is 2/0AWG (133100 cmil) copper. The soil has a uniform resistivity of 100 Ω·m. It is required

(a) to calculate the resistance to ground of the mat;
(b) to determine the effect of mat area and burial depth on the mat resistance to ground.

[2] J.G.Sverak, Simplified analysis of electrical gradients above a ground grid; Part I—How good is the present IEEE method? *IEEE Trans.Power Apparatus Syst.* vol. PAS-103, no. 1, pp.7–25, 1984.

(a) We have that $\sqrt{A} = 60\,m$, $\rho = 100\ \Omega\cdot m$, $h = 0.3\,m$
Total number of conductors = 32
Total length of wire $L = 32 \times 60 = 1{,}920\,m$.
From Eq.(6.2) the mat resistance to ground $R = 0.7384$
From Eq.(6.3) the mat resistance to ground $R = 0.7890$
From Eq.(6.4) the mat resistance to ground $R = 0.7970$

The mat resistance to ground for different values of mat area and burial depth has been calculated from Eq.(6.4) and the results are plotted in Figures 6.4 and 6.5. The conclusions are as follows:

- The resistance varies almost inversely as the square root of the mat area.
- The resistance decreases slightly with increasing burial depth.

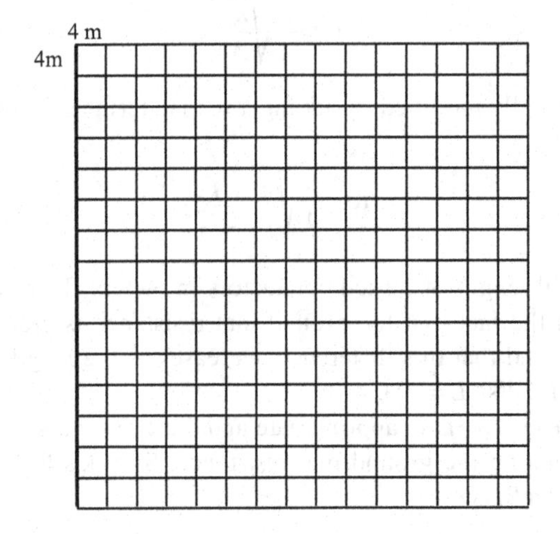

FIGURE 6.3 Ground mat divided into meshes 4m × 4m.

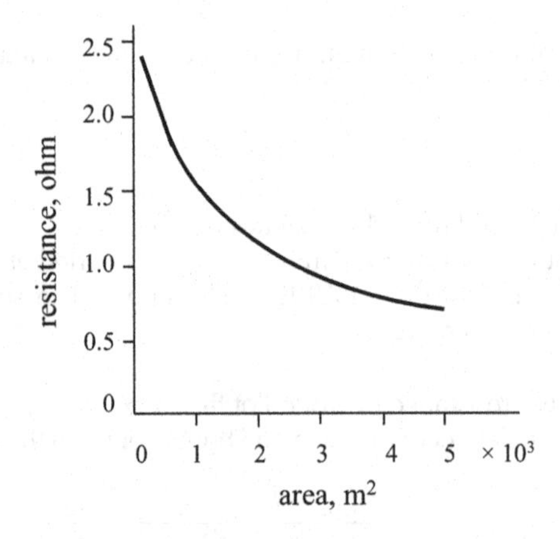

FIGURE 6.4 Effect of area on mat resistance.

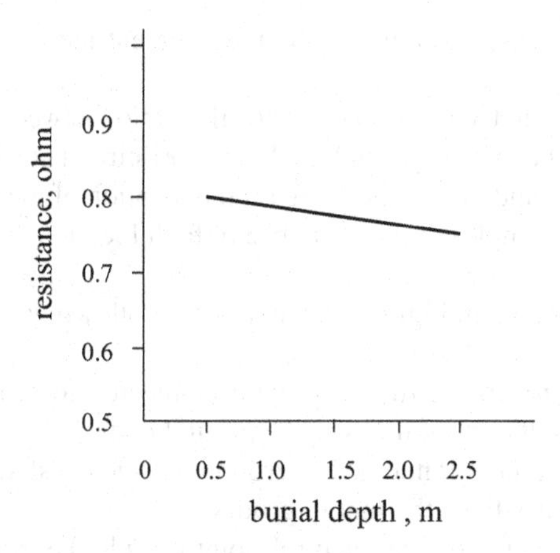

FIGURE 6.5 Effect of burial depth on mat resistance.

6.2.2 Calculation of Maximum GPR

To ensure the safety of personnel the ground mat must be designed so as to limit the GPR of the substation ground to an acceptable value when a short circuit to ground occurs. Determination of this GPR is quite complicated because it depends on many factors which include type of substation, location of fault, magnitude of fault current, and grounding system resistance. The maximum GPR can be calculated as follows:

$$(\text{GPR})_{\max} = I_G R_g \tag{6.5}$$

where

R_g = resistance of mat to ground.

I_G = it is the greatest value of rms current which flows from the ground grid to the surrounding ground. It is this current which produces the greatest rise in ground potential. The value of I_G is defined as follows:

$$I_G = C_f D_f S_f I_f \tag{6.6}$$

where

I_f = rms value of the symmetrical ground fault current;

C_f = correction factor for future system growth (>1);

D_f = decrement factor which takes into consideration the attenuation due to the passage of fault current from an asymmetrical to a symmetrical current:

Time, s	D_f
0.01	1.65
0.1	1.25
0.25	1.10
≥ 0.5	1.00

For intermediate values of fault duration, decrement factors may be obtained by linear interpolation.

S_f = current division factor. It represents the ratio between the current which actually flows from the grid to ground, and the short-circuit current and its value are usually between 50% and 80% depending on the location of the fault.

Figure 6.6 gives examples of the influence of fault location on the flow path of the fault current.

In the examples shown in Figure 6.6 we note the following:

In (a) the station neutral is the only point connected to ground. Almost the whole of the fault current flows in the ground mat.

In (b) the neutral point is connected to ground outside the station and the entire fault current flows from the mat to ground.

In (c) there is another grounded neutral point outside the station. One part of the fault current flows to ground.

In (d) the fault current divides between the grounded points according to the ground path resistances.

6.2.3 Determination of Maximum Touch and Step Voltages

The two most important parameters in the design of a substation ground mat are the touch and step voltages. Touch voltage in a grid ground is defined as the difference in potential between the grounded structure and the center of a grid mesh. The highest value of this voltage appears at the center of a mesh at a perimeter corner and is referred to as the "mesh voltage."

The maximum touch (mesh) and step voltages can be determined using the following formulas,[3]

(a) Grid without ground rods:

$$E_{mesh} = \rho K_m K_i I_G / L \tag{6.7}$$

$$E_{step} = \rho K_s K_i I_G / L \tag{6.8}$$

(b) Ground rods uniformly spaced within grounded area:

$$E_{mesh} = \rho K_m K_i I_G / (L + L_r) \tag{6.9}$$

$$E_{step} = \rho K_s K_i I_G / (L + L_r) \tag{6.10}$$

(c) Ground rods at the perimeter:

$$E_{mesh} = \rho K_m K_i I_G / (L + 1.15 L_r) \tag{6.11}$$

$$E_{step} = \rho K_s K_i I_G / (L + 1.15\ L_r) \tag{6.12}$$

[3] For a mathematical analysis of the gradient problem in a ground grid the reader should consult IEEE Std 80-2000: *Guide for safety in AC substation grounding*.

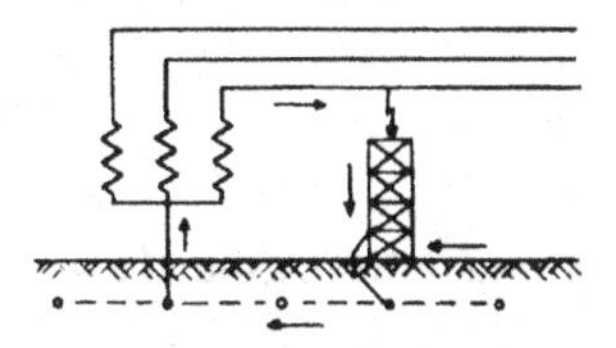

(a) Fault current follows metallic path provided by ground grid. No appreciable current flow in earth.

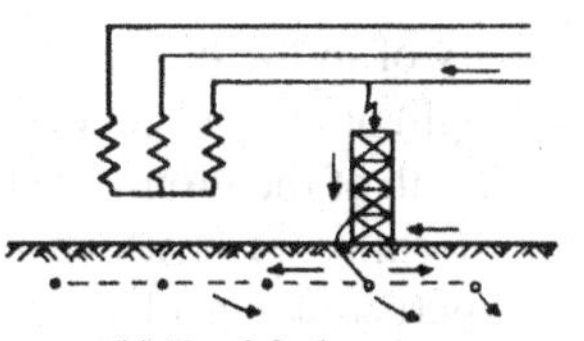

(b) Total fault current flows from station ground grid to earth.

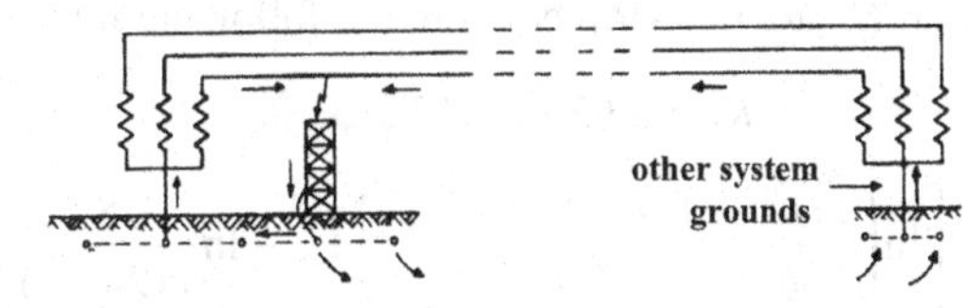

(c) Fault current returns to local neutral through local grid and to remote neutrals through earth. The latter current is of concern in study of danger voltages.

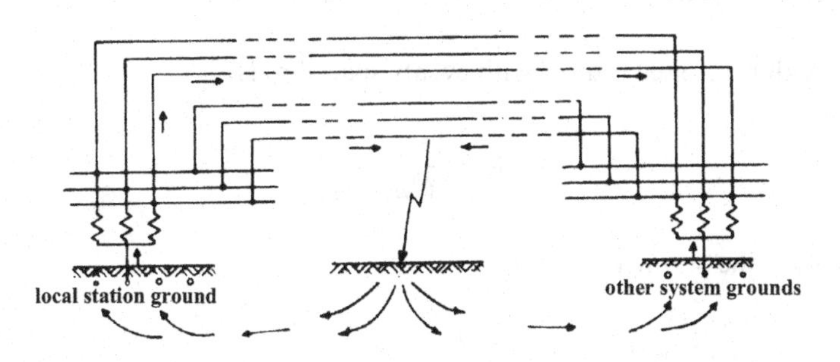

(d) Portion of fault current returning from earth to ground grid at local station determines grid potential rise and gradients there.

FIGURE 6.6 Fault location and path of current flow: (a), (b), and (c) fault within local station; (d) fault outside station. (Reproduced with permission from IEEE Std. 80 Guide for Safety in AC Substation Grounding.)

(a) Fault between a metal structure within the substation and a phase conductor of a three-phase line with the neutral point of the source grounded to the station mat. Most of the current flows through the ground mat and back up the neutral.
(b) Same as (a) but supply source neutral point not grounded. The fault current flows from grid to earth.
(c) Same as (a) but the station is connected to a second similar substation further down the line.One part of the current returns through the ground mat and the local neutral and the other part returns through earth and the neutral of the remote station.
(d) Two distant substations with grounded neural points are interconnected by two parallel three-phase transmission circuits. There is a fault between the phase conductor of one circuit and ground somewhere between the two stations. One part of the current flows through earth,ground mat, and neutral of one station and the remaining part flows through earth,ground mat, and neutral of the remote station.

where

ρ = resistivity of soil in ohm-m.

L = total length of all grid conductors (m).

L_r = total length of the ground rods (m).

K_i = nonuniformity factor which accounts for the nonuniform ground current flow from different parts of the grid.

K_s, K_m = coefficient which take into account the effect of number, spacing, size, and depth of burial of the grid conductors.

The values of K_i, K_s, and K_m are determined from the following formulas:

$$K_i = 0.656 + 0.172n \tag{6.13}$$

$$K_m = \frac{1}{2\pi}\left[\ln\left(\frac{D^2}{16hd} + \frac{(D+2h)^2}{8Dd} - \frac{h}{4d}\right) + \frac{K_{ii}}{K_h}\ln\frac{8}{\pi(2n-1)}\right] \tag{6.14}$$

$$K_s = \frac{1}{\pi}\left[\frac{1}{2h} + \frac{1}{D+h} + \frac{1}{D}(1-0.5^{n-2})\right] (0.25\text{m} < h < 2.5\text{m}) \tag{6.15}$$

For grids with no ground rods or with evenly spaced rods,

$$K_{ii} = \frac{1}{(2n)^{2/n}} \tag{6.16}$$

For rods along the perimeter,

$$K_{ii} = 1$$

For all cases,

$$K_h = \sqrt{1 + h/h_0} \tag{6.17}$$

n = number of parallel grid conductors in one direction,

D = spacing between parallel conductors (m),

h = burial depth (m),

h_o = 1 m (reference depth).

Figure 6.7 shows the voltage profile on the surface of the ground along a diagonal line above a ground mat with a 10 × 10 mesh and mesh size 6 m × 6 m. If a person standing at point P, which is at the center of the corner mesh, is in touch with a grounded part during the occurrence of a ground fault, he will be subjected to the touch voltage V_{touch}; a person standing at a corner with his feet 1 meter apart (F_1F_2) will be subjected to a step voltage V_{step} as shown.

Figure 6.8 shows the effect which the number of meshes into which a grid is divided has on the touch voltage. It can be seen that the maximum touch voltage (mesh voltage) decreases as the number of meshes increases.

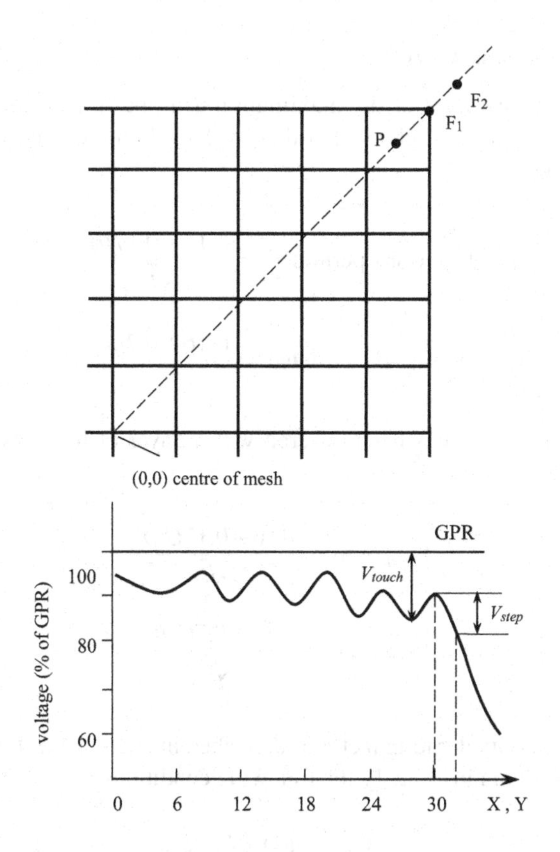

FIGURE 6.7 Voltage profile along the diagonal of a ground mat (10 × 10 mesh, size 6 m × 6 m, 2/0 AWG copper conductors, burial depth 0.3 m).

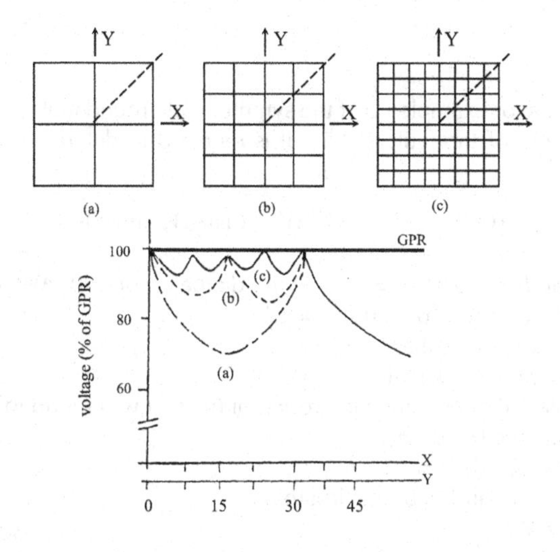

FIGURE 6.8 Effect of mesh size on touch voltage.

6.2.4 Safety Requirements

After calculating the values of the maximum touch (mesh) and step voltages from Eqs. (6.7) to (6.12) we must check that these values do not exceed those permitted (see Chapter 4) viz.,

$$V_{touch}(\text{max.permitted}) = \frac{(116+0.17\rho)}{\sqrt{t}} \tag{6.18}$$

$$V_{step}(\text{max.permitted}) = \frac{(116+0.7\rho)}{\sqrt{t}} \tag{6.19}$$

or if the ground surface has been covered with a layer of high resistivity such as crushed rock,

$$V_{touch}(\text{max}) = \frac{116+0.17C_s\rho_s}{\sqrt{t}} \tag{6.18a}$$

$$V_{step}(\text{max}) = \frac{116+0.7C_s\rho_s}{\sqrt{t}} \tag{6.19a}$$

where C_s is a resistivity derating factor as described in Section 4.2. Thus the safety of the grounding system must satisfy the following conditions:

$$V_{touch}(\text{max}) \geq E_{mesh} \tag{6.20}$$

$$V_{step}(\text{max}) \geq E_{step} \tag{6.21}$$

Example 6.2

For the ground mat of Example 6.1 if the symmetrical ground fault current is 1,000 A and the duration of the fault is 0.5 s, it is required to determine the following voltages:

$$(GPR)_{\text{max}}, E_{mesh}, E_{step}, V_{touch}(\text{max}), V_{step}(\text{max})$$

Depth of burial $h = 0.3$ m, $\rho = 100$ Ω·m, diameter of 2/0 AWG conductors $d = 0.0106$ m. Assume $S_f = 0.8$ and $C_f = 1.5$.

From Figure 6.3 we see that

$n = 16$, $D = 4$ m, $L = 1920$ m

From Example 1 the resistance to ground of the mat was found to be $R_g = 0.79$ Ω so from Eq.(6.5) we have that

$(\text{GPR})_{\text{max}} = 1200 \times 0.79 = 947$ V

and from Eqs.(6.7) and (6.8) we find that

$E_{mesh} = 152.24$ V

$E_{step} = 145.7$ V

and from Eqs. (4.4) and (4.3) we find that the maximum permissible touch and step voltages are

V_t(max) = 189 V

V_s(max) = 262 V

Since the two conditions (6.20) and (6.21) are met, no modifications to the mesh are necessary.

Example 6.3

Assume that the mesh size of the mat given in Example 6.2 is 10 × 10 m and all other details remain the same. It is required to find the touch and step voltages and introduce any necessary modifications required to make these voltages safe.

From the appropriate equations we find that

$$E_{mesh} = 295\,\text{V}, E_{step} = 158\,\text{V}$$

$$V_t(\text{max}) = 189\,\text{V}, V_s(\text{max}) = 262\,\text{V}$$

The touch voltage does not satisfy the safety requirement. It is possible to introduce two modifications in the design:

(a) Decrease E_{mesh} and E_{step} by increasing the length of buried conductors using a number of rods driven into the ground mat area:
number of rods = 49
length of each rod = 15 m
rod diameter of rod = 1.9 cm
total length of rods L_r = 735 m
From Eqs.(6.9) and (6.10) we find that

$$E_{mesh} = 157\,\text{V} < V_t(\text{max}) = 189\,\text{V}$$

$$E_{step} = 84\,\text{V} < V_s(\text{max}) = 262\,\text{V}$$

(b) It is possible to raise the maximum permissible values of the touch and step voltages [V_s(max), V_t(max)] by using a layer of crushed rock:
resistivity ρ_s = 2,500 Ω·m
thickness h_s = 0.1 m
From Eqs. (6.18a) and (6.19a) we find that

$$V_t(\text{max}) = 465\,\text{V} > E_{mesh} = 295\,\text{V}$$

$$V_s(\text{max}) = 1{,}402\,\text{V} > E_{step} = 158\,\text{V}$$

The correction factor C_s used in Eqs.(6.18a) and (6.19a) is taken from Figure 4.3 for $h_s = 0.1\text{m}$ and $K = -0.9$.

6.3 DESIGN OF GROUND MAT USING COMPUTER PROGRAM

The reader will realize from the foregoing that the design of ground mats requires laborious calculations. However, there are a number of computer programs specially designed for this purpose and we give here an example of mat design using one such program (EDSA) which uses the Sverak formula as given by Eq. (6.4).

Figure 6.9 shows the preliminary design of a substation ground mat of area 150 × 150 ft. The size of the copper conductors is 2/0 AWG and the mat is buried at a depth of 1.5 ft in soil of uniform resistivity 100 ohm-m. The fault current is 1,000 A and its duration 0.25 s. It is required to determine from the program if the safety requirements are met in the following cases:

(A) If the mesh size is 15 × 15 ft.
(B) If the mesh size is 30 × 30 ft.
(C) If the surface is covered with a layer of crushed rock of resistivity 2,500 Ω.m and thickness 4″.

Tables 6.1–6.3 give the respective computer outputs for these three cases:

(A) Table 6.1 shows that safety requirement are met since E_{mesh} and E_{step} are less than V_t(max) and V_s(max).
(B) Table 6.2 shows that 20 driven rods are required to meet the safety requirements.
(C) Table 6.3 shows the large increase in V_t(max) and V_s(max) brought about by the using a layer of crushed rock.

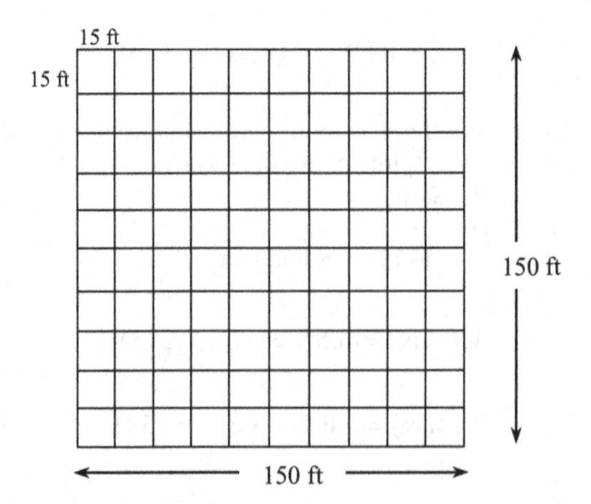

FIGURE 6.9 Ground mat of area 150 ft^2 with 10 × 10 meshes, burial depth 1.5 ft.

TABLE 6.1

Job: Solution of Example (Case A) Ground Grid Design using EDSA Program EE Dept. College of Engineering

Grid Data

Grid length	150.00ft	Grid width	150.00ft
Conductor spacing	15.00 ft	Ambient temperature	77.00 F
Sym. grnd. fault	1,000 amps	Fault duration	0.25 s

Current Division Factor 0.80

Resistivity

Soil	100.00 Ω·m	Crushed rock	2,500.00 Ω·m
Concrete	0.00 Ω·m		

Depth of Burial

Grounding grid	1.50 ft	Crushed rock	0.00 ft
Reference depth of grid	3.28 ft	Concrete radius	0.00 ft

Ground Rod Information

Number	Length (ft)	Diameter (in.)
20	24.61	0.748
36	32.81	0.748
36	49.21	0.748
38	32.81	0.748
48	39.37	0.748

Material	Conductivity	Factor	Oc	FusingTemp.	Resistivity	Thermal Capacity
STDANNEALED-SOFT-CU	100.00	0.003930	234.0	1,083.0	1.724	3.422

AWG/MCM	Cir.Mil	Cond.Dia CM	Con.Dia IN
2/0 AWG	133,100	1.0600	0.4190

Output Results			
Voltages			
V (max. touch)	268.33	V (max. step)	377.32
E (mesh)	171.06	E (step)	107.19

Grid potential rise 929.84

Grid Particulars	
Conductor length	3,300.00 ft
Max grid current	880.00 amps
Grid resist	1.057 Ω

Material	AWG/MCM	Cirmil Grid	Cond Dia (in.)
STD-ANNEALED-SOFT-CU	2/0 AWG	133,100	0.4173

*** No rods are necessary ***

TABLE 6.2

Job: Solution of Example (Case B) Ground Grid Design using EDSA Program EE Dept. College of Engineering

Grid Data:

Grid length	150.00	ft	Grid width	150.00 ft
Conductor spacing	30.00	ft	Ambient temperature	77.00 F
Sym. grnd. fault	1,000	amps	Fault duration	0.25 s

Current division factor 0.80

Resistivity:			
Soil	100.00 Ω·m	Crushed rock	2,500.00 Ω·m
Concrete	0.00 Ω·m		

Depth of burial:			
Grounding grid	1.50 ft	Crushed Rock	0.00 ft
Reference depth of grid	3.28 ft	Concrete Radius	0.00 ft

Ground Rod Information		
Number	**Length (ft)**	**Diameter (in.)**
20	24.61	0.748
36	32.81	0.748
36	49.21	0.748
38	32.81	0,748
48	39.37	0.748

Material	**Conductivity**	**Factor**	**Oc**	**Fusing Temp**	**Resistivity**	**Thermal Capacity**
STD-ANNEALED-SOFT-CU	100.00	0.003930	234.0	1,083.0	1.724	3.422

AWG/MCM	**Cir.Mil**	**Cond.Dia CM**	**Con.Dia IN**
2/0 AWG	133,100	1.0600	0.4190

Output Results			
Voltages			
V (max. touch)	268.33	V(max. step)	377.32
E (mesh)	192	E (step)	85.25
Grid potential rise	1,095.16		

Grid Particulars	
Conductor length	1,800.00 ft
Max grid current	880.00 amps
Grid resist	1.244 Ω

Material	AWG/ MCM	Cirmil	Grid Cond Dia (in)	Ground Rods	Length of Each Rod	Dia.of Rod
STD-ANNEALED-SOFT-CU	2/0 AWG	133,100	0.4173	20	24.61 ft	0.748 in.

TABLE 6.3

Job: Solution of Example (Case C) Ground Grid Design using EDSA Program EE Dept. College of Engineering

GRID DATA:

Grid length	150.00 ft	Grid width	150.00 ft
Conductor spacing	15.00 ft	Ambient temperature	77.00 F
Sym. grnd. fault	1,000 amps	Fault duration	0.25 s

Current division factor 0.80

Resistivity:			
Soil	100.00 Ω·m	Crushed rock	2,500.00 Ω·m
Concrete	0.00 Ω·m		

Depth of Burial:			
Grounding grid	1.50 ft	Crushed rock	0.33 ft
Reference depth of grid	3.28 ft	Concrete radius	0.00 ft

Ground Rod Information		
Number	Length (ft)	Diameter (in)
20	24.61	0.748
36	32.81	0.748
36	49.21	0.748
38	32 81	0.748
48	39.37	0.748

Material	Conductivity	Factor	Oc	Fusing Temp	Resistivity	Thermal Capacity
STD-ANNEALED-SOFT-CU	100.00	0.003930	234.0	1,083.0	1.724	3.422

AWG/MCM	Cir.Mil	Cond.Dia. CM	Con.Dia. IN
2/0 AWG	133,100	1.0600	0.4190

Output Results			
Voltages			
V (max. touch)	712.58	V (max. step)	2,154.31
E (mesh)	155.51	E (step)	97.44
Grid potential rise	845.31		

Grid Particulars	
Conductor length	3,300.00 ft
Max grid current	800.00 amps
Grid resist	1.057 ohms

Material	AWG/MCM	Cirmil	Cond Dia (in)
STD-ANNEALED-SOFT-CU	2/0 AWG	133,100	0.4173

*** No rods are necessary ***

7 Static Electrification

7.1 INTRODUCTION

Static electrification is a term applied to all mechanical operations which result in the separation of positive and negative charges. These operations include friction, contact, or impact between two solid surfaces, a solid surface and a liquid or gas as well as the separation of surfaces and the rupture of liquids by spraying or bubbling. Electrification by friction or contact is generally known as *tribo-electrification* or *contact electrification* and the static electrification produced by spraying (atomizing) or bubbling of liquids is known as *spray electrification*. The electrification of dusts and powders is a type of solid-solid tribo- or contact-electrification. It should be mentioned here that experimental results have confirmed that tribo-electrification is an effect due to contact between two surfaces and their subsequent separation; friction has no effect on the electrification process unless it leads to an increase in the temperature of one surface above that of the other. The phenomena of static electrification are of major importance in industry and can cause explosions in sugar factories, granaries, sulfur mills, explosives factories, petrochemical factories, and in all branches of the petroleum industry as well as in operations involving the handling of coal and inflammable liquids. Moreover, static electrification has been known to cause explosions in the operating rooms in hospitals.

With the great advances in the manufacture of integrated circuits and their use in all modern electronic devices and systems, electrostatic discharge (ESD) constitutes a serious source of damage to these circuits during their manufacture or their packaging, as well as during their operation. Experience has shown that such discharges lead to the destruction of thin film metal oxide semiconductors and a number of sensitive elements such as tracks, membrane resistances, and capacitors in integrated circuits, etc. They may also cause the so-called ESD latent static damage in which an initial discharge will cause only partial damage to a track leading to its gradual degradation and eventual failure due to thermal or vibrational stresses during use.

Before considering the subject of static electrification and ways to control it, we have found it pertinent to acquaint the reader with a synopsis of the conditions which are necessary for an ESD to cause a fire or an explosion.

7.2 CONDITIONS NECESSARY FOR IGNITION

Burning of a fuel in air is the result of the reaction between the fuel molecules and the oxygen molecules in the air. For liquid fuels the mixture of fuel and oxygen becomes available when the fuel evaporates, but for solid fuels there must be a break in the chemical bond of the fuel molecules so that the resulting hydrocarbon molecules can interact with the oxygen.

A fire or explosion will occur when the following three elements exist together: fuel, oxygen, and an ignition source; these constitute the so-called danger triangle. However, the simultaneous presence of these elements per se will not lead to a fire or explosion unless the following conditions prevail:

(a) The composition of the combustible mixture is within the explosion limits (EL) of the fuel.
(b) The concentration of oxygen is not below 10%.
(c) The energy released by the ignition source is at least equal to the minimum ignition energy (MIE) of the mixture.

7.2.1 Explosion Limits

For any kind of fuel, whether gaseous, liquid, or solid, self-sustained combustion is only possible for volume concentrations between a lower and upper limit known as the explosion or flammability limits (EL or FL) which must be determined by experiment.

The lower explosive limit (LEL) is the lowest concentration of a gas or vapor in air which will burn or explode in the presence of an ignition source. Below the LEL the mixture is "too lean" to burn (i.e., there is insufficient fuel). The upper explosive limit (UEL) is the highest concentration of a gas or vapor in air which will burn or explode in the presence of an ignition source. Above the UEL the mixture is "too rich" to burn (i.e., there is insufficient oxygen).

For gases and vapors both these concentrations are expressed as the percentage (volume) concentration of fuel in the fuel/air mixture (in oxygen-rich atmospheres, such as hospitals, concentrations should be known for fuel/oxygen mixtures). For dusts (or powders) they are expressed as the mass of dust per unit volume of dust/air mixture, e.g., kg/m^3.

(i) Liquid and gaseous fuels.

The ignition of the liquid fuel depends on its vapor pressure. This pressure, and hence the vapor concentration, depends on the temperature. From the relationship between vapor pressure and temperature for a given liquid we obtain the relationship between vapor concentration and temperature. The temperature corresponding to the LEL is known as the *flash point* of the liquid, which is defined as the lowest temperature at which the vapor concentration in the air near the surface of the liquid is high enough to form an ignitable mixture. The flash point of a liquid is a measure of its flammability, the lower its flash point, and the higher its flammability. At temperatures below their flash point liquids cannot be ignited.

Tables 7.1 and 7.2 give the explosive limits, the flash point, and the MIE (Sec.7.2.4) for a number of substances. The values have been chosen from a large number of sources [38–40] and may be finely adjusted from time to time. However, they serve as guidance and as a comparison between the FLs and MIE of some common gases and vapors.

TABLE 7.1
LEL and UEL and Flash Point of Some Gases and Vapors

Substance	LEL (% by vol of air)	UEL (% by vol of air)	Flash Point (°C)
Acetone	2.6–3	12.8–13	−17
Acetylene	2.5	82	−18
Benzene	1.2	7.8	−11
Butane	1.4	8.4	−60
Carbon monoxide	12	75	−191
Diesel fuel	0.6	7.5	> 62
Diethyl ether	1.9–2	36–48	−45
Ethanol (ethyl alcohol)	3–3.3	19	12.8
Ethylene glycol	3	22	111
Gasoline (100 octane)	1.4	7.6	< −40
Hexane	1.1	7.5	−22
Hydrogen	4	75	–
Isopropyl alcohol	2	12	12
Methane	4.4–5	15–17	–
Methanol (methyl alcohol)	6–6.7	36	11
Octane	1	7	13
Pentane	1.5	7.8	− 40 to −49
Propane	2.1	9.5–10.1	–
Toluene	1.25	6.75–7.1	4.4
Xylene	0.9–1.0	0.9–1.0	27–32

TABLE 7.2
MIE for Some Gases and Vapors

Substance	MIE (mJ)	Concentration in Air (% by volume)
Acetone	1.15	4.5
Acetylene	0.017	8.5
Ammonia	680	–
Benzene	0.22	4.7
Butane	0.25	4.7
Diethyl ether	0.19	5.1
Hexane	0.24	3.8
Hydrogen	0.017	28
Methane	0.28	8.5
Methanol	0.14	–
Propane	0.25	5.2
Toluene	0.24	4.1

(ii) Dust layers and dust clouds

Dusts and powders also have lower and upper ELs but in general it is the lower limits which are of importance. For many organic dusts these are in the range 10–50 g/m^3, which is much higher than the limit set for health reasons. ELs also depend on the size of the dust particles involved and are not an intrinsic property of the material. A concentration above the LEL can be created from an accumulation of settled dust by a sudden gust of wind. Dust accumulations can be prevented by using appropriate enclosures, proper ventilation, and regular surface cleaning. However, the formation of dust clouds is inherent in certain industries such as sugar mills, sulfur mills, flour, and grain industries as well as in granaries and in the handling of coal. Such dust clouds are explosive and some can be ignited by static sparks.

7.2.2 Oxygen

Oxygen is a vital element in the combustion of any type of fuel and the majority of fuels require a minimum oxygen concentration of about 10% by volume for combustion to take place. Since under normal atmospheric conditions the oxygen content of air is 21% by volume, this means that for absolute safety about 50% of the air must be substituted by some inert gas such as nitrogen or carbon dioxide. This may be implemented only in certain limited zones where it is absolutely necessary to guarantee a 100% explosion proof environment.

7.2.3 Ignition Sources

An ignition source is a source that is capable of providing the energy necessary to ignite a combustible mixture. The possible sources of ignition are the following (not listed in order of importance or of frequency of occurrence):

- Flames (cutting and welding, cigarettes, matches, etc.)
- Hot surfaces (heating pipes, electrical casings, etc.)
- Friction heating or sparks (abrasive cutting)
- Mechanical machinery
- Electrical equipment and installations (electrical sparks at make/break switches, loose contacts, bad solder joints in cables, etc.)
- ESD sparks
- Lightning strikes
- Electromagnetic radiation of different wavelengths
- Ultrasonics
- Exothermal chemical reactions

The potential of these sources to ignite a combustible mixture varies enormously. For example, whereas lightning and an open flame have sufficient energy to ignite any combustible material, other sources such as mechanical or electrical sparks can only do so if their discharge energy is equal to or greater than the MIE of the mixture.

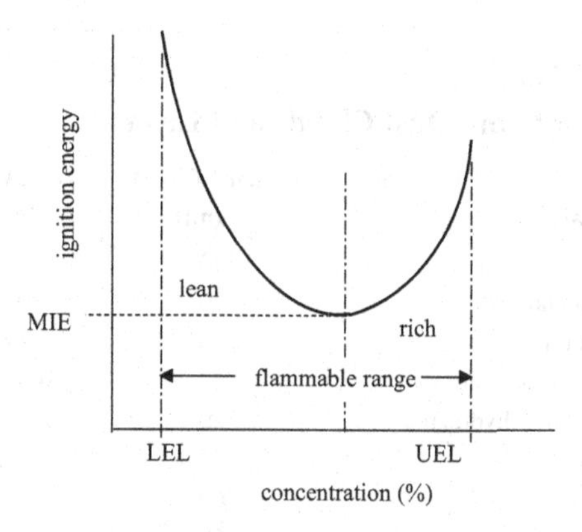

FIGURE 7.1 Variation of ignition energy with concentration of combustible material.

7.2.4 Minimum Ignition Energy

As mentioned above, for ignition to occur, the concentration of the combustible material must lie between the lower and upper explosion limits. For each concentration within these limits a certain amount of energy (usually measured in milli-joules) is required to initiate the ignition process. As we move from LEL to UEL concentrations this energy passes through a minimum value as shown in Figure 7.1. This value represents the minimum ignition energy (MIE of the mixture. Table 7.2 gives the MIE values for some vapors and gases whereas Table 7.3 gives MIE values for dust clouds and dust layers.

7.3 GENERATION OF ELECTROSTATIC CHARGES

Surface charges are generated when two dissimilar surfaces are brought together and then separated. During the initial contact a transfer of charge takes place between the surfaces so that when they are separated one surface will carry a negative charge and the other a positive charge. This method of static charge generation is known as tribo-electrification although, as mentioned in the introduction, it is in effect contact electrification.

The amount of charge that is transferred from one body to another depends on the ability of the material that donates charge or its ability to accept charge. Based on this ability materials have been classified in a series known as the *tribo-electric series* which is given in Table 7.4. Materials whose position in the table is higher than others have a greater ability to donate electrons (and hence acquire a positive charge) than those lying below it. Similarly materials whose position is lower than others have a greater ability to accept electrons (and hence acquire a negative charge) than those above them in the table. Thus if two bodies of dissimilar material come into contact the body which acquires a positive charge is that whose position in the table is above that of the other. For instance if a glass body comes in contact with a rubber body, the glass will acquire a positive charge and the rubber a negative charge.

The magnitude of the acquired charge depends on a number of factors such as the cleanliness of the surfaces, contact pressure, extent of contact area as well as the speed with which the surfaces are separated. In many cases surface charges may

TABLE 7.3
MIE for Some Dust Clouds and Layers

Material	Dust Cloud (mJ)	Dust Layer (mJ)
Alfalfa	320	
Allyl alcohol resin	20	80.0
Aluminum	10	1.6
Aluminum stearate	10	40.0
Aryl sulfonyl hydrazine	20	160.0
Aspirin	25	160.0
Boron	60	
Cellucotton	60	
Cellulose acetate	10	
Cinnamon	40	
Coal, bituminous	60	560.0
Cocoa	100	
Cork	35	
Cornstarch	30	
Dimethyl terephthalate	20	
Dinitro-o-toluamide	15	24.0
Ferromanganese	80	8.0
Gilsonite	25	4.0
Grain	30	
Hexamethylenetetramine	10	
Iron	20	7.0
Magnesium	20	0.24
Manganese	80	3.2
Methyl methacrylate	15	
Nut shell	50	
Paraformaldehyde	20	
Pentaerythritol	10	
Phenolic resin	10	40.0
Phthalic anhydride	15	
Pitch	20	6.0
Polyethylene	30	
Polystyrene	15	
Rice	40	
Seed (clover)	40	
Silicon	80	2.4
Soap	60	3,840.0
Soybean	50	40.0
Stearic acid	25	
Sugar	30	
Sulfur	15	1.6
Thorium	5	0.004
Titanium	10	0.008
Uranium	45	0.004

(Continued)

TABLE 7.3 (*Continued*)
MIE for Some Dust Clouds and Layers

Material	Dust Cloud (mJ)	Dust Layer (mJ)
Urea resin	80	
Vanadium	60	8.0
Vinyl resin	10	
Wheat flour	50	
Wood flour	20	
Zinc	100	400.0
Zirconium	5	0.0004

Source: Data from the US Bureau of Mines. Reprinted with permission from IEEE Std. 142-2007.

TABLE 7.4
The Tribo-Electric Series

Positive					
1	Air	13	Paper	25	Acrylic
2	Human skin	14	Cotton	26	Polyester
3	Asbestos	15	Wood	27	Celluloid
4	Glass	16	Steel	28	Orlon
5	Mica	17	Sealing wax	29	Polyurethane
6	Human hair	18	Hard rubber	30	Polyethylene
7	Nylon	19	Mylar	31	Polypropylene
8	Wool	20	Epoxy	32	PVC
9	Fur	21	Copper, nickel	33	Silicon
10	Lead	22	Silver, brass	34	Teflon
11	Silk	23	Gold, platinum		**Negative**
12	Aluminum	24	Spongy polystyrene		

appear when two similar materials are separated as often happens when plastic bags are opened or a sheet of "cling film" is drawn from a roll. In such cases the surface charging is attributed to the transfer of ions due to the presence of impurities.

7.3.1 Contact Electrification between Solid Surfaces

(A) Contact between two dissimilar metals

In this case free electrons move from the surface of the metal with the lower work function[1] to the one with the higher work function. As a result

[1] The work function is the energy required to remove a free electron from the surface of a pure metal in vacuum and is measured in electron-volt (eV).

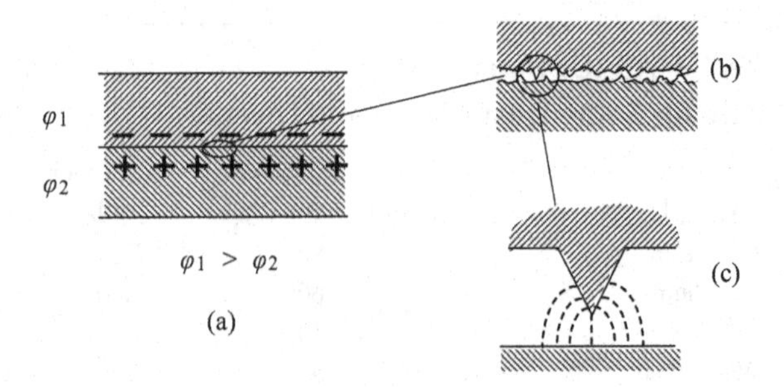

FIGURE 7.2 (a) Formation of surfaces charges due to differences in work function φ; (b) surface roughness; (c) enhancement of electric field at sharp points.

of this transfer there will be, at equilibrium, a layer of negative charges on one of the surfaces and a layer of equal positive charges on the other surface as shown in Figure 7.2a.

(B) Contact between metal and insulator

In this case too, positive and negative charges are formed over the surfaces although not because of a difference in work function but because of the transfer of electrons (possibly by tunneling) as well as of positive and negative ions from one of the surfaces to the other. However, the exact mechanism of transfer is not well defined [14].

(C) Contact between insulating surfaces

The transfer of charge in this case depends on the basic structure of the insulating materials and the type and amount of ionic contamination on the surfaces and in the atmosphere, as well as on the densities of donors and acceptor impurities present in each material.

(D) Separation of surfaces

When two charged surfaces are separated the capacitance between them decreases, and this is accompanied by a large increase in the potential difference since the charge remains constant in the relationship

$$V = Q/C$$

The external work done to separate the surfaces appears as energy stored in the electrostatic field ($\frac{1}{2}QV$). In practice, however, the separation of two surfaces is accompanied by a reduction in surface charge. There are two reasons for this:

The first reason is that on a microscopic scale surfaces are far from smooth (Figure 7.2b) so that during the separation process charges tend to accumulate at the last remaining points of contact between the two surfaces. Because such points can be quite sharp they produce a field of very high intensity in the gap (Figure 7.2c) leading to the discharge of surface charges through such points by field emission, a process referred to as field emission back discharge. The speed of this back discharge will depend upon the speed with which the charges move to the discharge points and this, for a given separation velocity, depends upon the surface resistivity. The lower this resistivity the smaller the charges left on the surfaces after separation. Also, the faster the separation the greater the number of charges that will be left on the surfaces.

The second reason for the reduction of surface charges on separation is that during the separation the field in the gap between the two surfaces (depending on the initial surface charge density) may attain a value of 30 kV/cm, which is the breakdown strength of air and is sufficient to produce ionization and initiate local discharges. The positive and negative ions resulting from such a process will be drawn to the surfaces and neutralize their opposite charges there.

Because of the above-mentioned reasons it is difficult to predict the amount of charge that will remain on surfaces after their separation. Table 7.5 gives some typical values for the electrostatic voltages which can be generated by tribo-electrification; the magnitude of these voltages is strongly dependent on the relative humidity (RH), but voltages of 10 and 20 kV are certainly possible.

7.3.2 Static Electrification in Liquids and Gases

When a metallic surface comes in contact with an electrolyte which has a high relative permittivity (such as water, $\varepsilon_r = 80$) an exchange of ions takes place between the two media leading to the formation at the interface of a layer of positive charges and a layer of negative charges known as the *Helmholtz double layer* as shown in Figure 7.3.

These charges become separated from each other when the liquid flows. For example, if the liquid flows in a metal pipe and is collected in a tank which is insulated from the pipe, charges accumulate in the tank and create an ignitable atmosphere. An example of the latter is the washing of empty oil tanks of tanker ships;

TABLE 7.5
Some Typical Statically-Generated Voltages

	Voltage(V)	
	Relative Humidity	
Means of Static Generation	**10%–20%**	**65%–90%**
Walking across carpet	35,000	1,500
Walking on vinyl floor	12,000	250
Worker moving at bench	6,000	100
Opening a vinyl envelope	7,000	600
Picking up common polyethylene bag	20,000	1,200
Sitting on chair packed with polyurethane foam	18,000	1,500

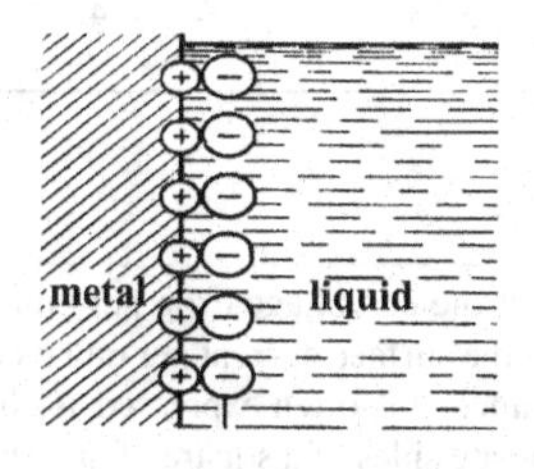

FIGURE 7.3 Formation of Helmholz double layer at metal/liquid interface.

these tanks always have a flammable atmosphere of hydrocarbon gases/air mixture. Spraying with statically charged water during washing creates a tribo-electrically charged mist which under certain favorable conditions can cause extensive ESDs within the tank which are known to have been the cause of explosions. This phenomenon, which is known as *flow electrification*, can occur with insulating liquids which have a high content of moisture or other electrolytic impurities.

Static electrification in liquids can also occur at the interface between two liquids, especially when this surface is very large as it is in the case of emulsions of oil (or any other hydrocarbon liquid) and water.

In the case of clean gases flowing in metal pipes there is no static charge generation. However, if the gases contain entrained dust particles or aerosols then charges may form.

7.4 WAYS FOR REDUCING THE FORMATION OF SURFACE CHARGES

It has been mentioned in Section 7.3.1 that the charge which appears on the surfaces of two metals during their contact depends essentially on the difference in their work functions. It is therefore possible to reduce surface charges by choosing metals with as small a difference as possible in their work functions. The work function of a number of metals is given in Table 7.6.

In the case of insulating surfaces the materials may be chosen so that they lie close to one another in the tribo-electric series, although this may not always be a guarantee. Perhaps the best method for minimizing the charges which appear on the separation of such surfaces is to reduce their surface resistivities. Experimental investigations [10] have shown that for a large number of plastic materials and a normal separation speed of 1 m/s, surface charges could only be detected on surfaces whose resistivities exceeded 10^{12} Ω/sq.[2]

TABLE 7.6
The Work Function (eV) of Pure Metals

Li	2.48	Be	3.32–3.92	Th.	3.38	Fe	4.49
Na	2.28	Mg	3.67	C	4.35–4.60	Ni	4.96
K	2.22	Ca	3.20–3.71	Si	3.54	Pd	4.98
Cs	1.93	Ba.	2.51	Ta	4.13	Pt	5.36
Cu	4.45	Zn	4.29	C	4.60		
Ag.	4.46	Al	4.20	M	4.24		
An.	4.89	Zr	3.73	W	4.54		

[2] The surface resistivity is the ratio of the dc voltage drop per unit distance between two parallel electrodes in contact with the surface to the surface current per unit electrode length. Its unit is ohms, but to avoid confusion with surface resistance, it is often expressed as ohm/square. It is numerically equal to the surface resistance between opposite sides of a square of any size when the current flow is uniform.

According to their surface resistivities materials are classified into three groups:

- > 10^{12} Ω/sq nondissipative
- 10^{10}–10^{12} Ω/sq antistatic (reduce the amount of charge generated by contact and separation)
- 10^{5}–10^{9} Ω/sq statically dissipative (antistatic)
- 100–10^{5} Ω/sq conductive (no charge accumulation)

The surface resistivity of some copolymer plastics can be reduced to 100–1,000 Ω/sq by the addition of fillers such as carbon powder, carbon fibers, or stainless steel fibers. Applications of conductive plastics include their use in medical devices, for packaging of electronic devices and for ESD protection, as conductive storage containers for hazardous liquids, and wherever protection from ESD is needed.

One family of copolymer plastics, e.g., polyanilines, have a low surface resistivity in the range of 100–1,000 Ω/sq and are known as inherently conducting polymers. They are used mainly in antistatic applications, batteries, and photochemical cells; their main drawback, so far, has been their processability and thermal stability.

Table 7.7 gives typical values of the surface resistivities of a number of commonly used plastic materials.

One of the most important factors which influence the surface resistivity of many materials is the relative humidity of the surrounding atmosphere and the capability of the material to absorb moisture. Surface resistivities should therefore be measured for the same relative humidity values as those expected in actual use. Some polymers, especially silicones, have a very low moisture absorbency and their surface resistivity is unaffected by humidity.

TABLE 7.7
Surface Resistivity of Some Plastics

Material	ρ_s(Ω/sq)
CA cellulose	10^{12}–10^{14}
FEP fluorinated ethylene propylene copolymer	10^{16}
PA 6 polyamide—nylon 6	5×10^{10}
PA 12 polyamide—nylon 12	10^{13}
PC polycarbonate	10^{15}
PC polycarbonate—conductive	100–500
PE polyethylene—carbon-filled	10^{3}–10^{4}
PE(HD) (polyethylene—high density)	10^{13}
PE(LD) (polyethylene—low density)	10^{13}
PMMA (polymethylmethacrylate)	10^{14}
PP (polypropylene)	10^{13}
PS (polystyrene—conductive)	10^{2}–10^{7}
XLPS (polystyrene—cross-linked)	>10^{15}
PTFE (polytetrafluoroethylene)	10^{17}
PVC (polyvinylchloride) soft	10^{13}
PVDF (polyvinylidenefluoride)	10^{13}

7.5 ELECTROSTATIC INDUCTION

Figure 7.4 shows how induced charges appear on a conducting body. When an isolated charged body carrying a negative charge for example is brought near the conductor, a positive charge appears on the surface of the conductor nearest the charged body and an equal negative charge will appear on the far surface since the total charge on the conductor is zero. If we assume that the conducting body is now momentarily connected to ground as shown in Figure 7.4a, the induced negative charges will flow to earth but the positive charges will remain unchanged since they are bound to the negative charges on the charged body (Figure 7.4b). If the isolated charged body is now removed from the vicinity of the conductor, the positive charges are now free to redistribute themselves over the surface (Figure 7.4c). We thus see that a conducting body can acquire a charge without making direct contact between it and any charged body.

When the conductor which has acquired an induced charge approaches another conducting body the whole or part of this charge will flow to ground via an arc discharge (Figure 7.4d). If the body is not grounded, the discharge current will flow to earth through the capacitance between this body and ground.

7.6 TYPES OF ELECTRIC DISCHARGES

There are basically three types of electrical discharges which occur in air (or any other gaseous medium):

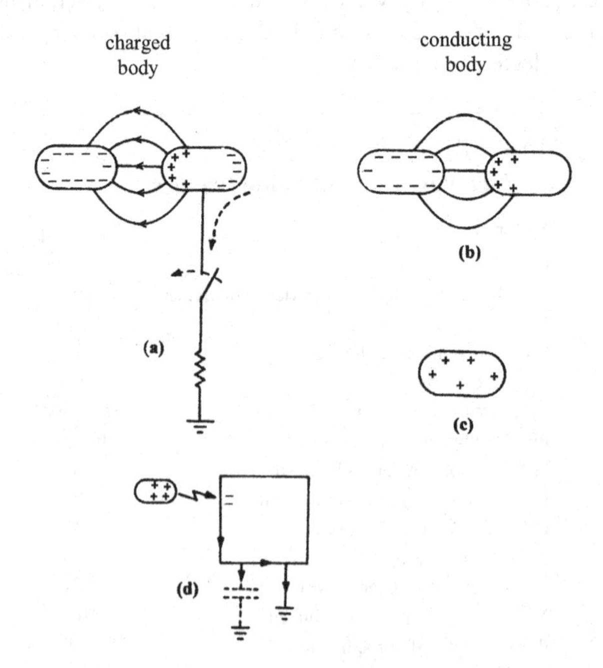

FIGURE 7.4 (a–c) Charging of a conducting body by induction; (d) ESD.

- Spark or arc discharges
- Brush and propagating brush discharges
- Corona discharges

The first two are potentially dangerous since they can cause damage to electronic equipment and explosions in combustible atmospheres, while corona discharges have too low an energy to cause ignition and are in fact used as a preventive measure against charge accumulation.

7.6.1 The Spark or Arc Discharge

When the electric field in the gap between two conductors reaches the breakdown strength of air (30 kV/cm at normal temperature and pressure) there is an electrical discharge which completely bridges the gap in the form of a spark (which initiates an arc if the electrodes are connected to a supply). The energy stored in the interelectrode capacitance ($\frac{1}{2}CV^2$) is dissipated in the spark discharge. If the discharge occurs in an explosive atmosphere ignition may result if the discharge energy is greater than the MIE of the medium. Of major concern too is the damage which such discharges can cause to sensitive electronic equipment. Sparks occur between charged conductors. Since human beings are electrically conducting they can carry large induced charges and as such are the prime source of spark discharges.

As a practical example suppose that a person wearing rubber-soled shoes is walking on a nylon or woolen carpet. From Table 7.4 we deduce that the soles will acquire a negative charge, and since the body is a conductor, a positive-induced charge will appear on the soles of the person's feet and a negative charge on the upper part of his body, especially at the tips of his fingers (Figure 7.5). As his hand approaches a grounded metallic body an ESD will take place between the fingers and the metallic body when the electric field between them exceeds the breakdown strength of air (30 kV/cm). Humans cannot feel a discharge of less than 3500 V and discharges at potentials greater than 25 kV are painful.

Because the human body is a conductor it acts as a capacitor in which charge can be stored. Now the absolute capacitance of any conductor is its capacitance as an

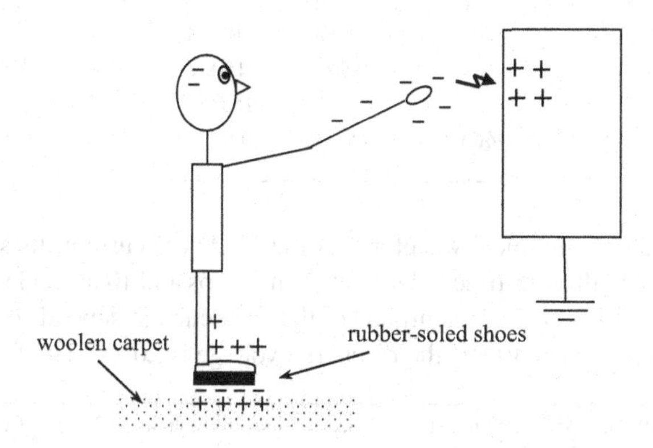

FIGURE 7.5 Discharge from person to metallic body.

isolated body, i.e., between the conductor and infinity. The simplest example is the capacitance of an isolated conducting sphere of radius r:

$$C = 4\pi\varepsilon_o r$$

$$C = 111 rpF$$

Since in general the capacitance of a body is a function of its surface area, it is possible to consider the human body as equivalent to a spherical conductor of diameter 1 m; thus the absolute capacitance of the human body is approximately 50 pF. Typical values are in the range 50–100 pF.

When other bodies are in the proximity of a conductor, there will be other capacitances between the conductor and these other bodies; these are the mutual capacitances. For the human body the mutual capacitances are those between the body and surrounding walls and those between the soles of the feet and ground (Figure 7.6). If all these capacitances are taken into consideration, the capacitance of the human body will be in the range of 50 pF to 250 pF.

As mentioned in Chapter 1 the resistance of the human body can vary widely from 10 kΩ if the discharge occurs from the tip of a finger to 1,000 Ω if from the palm of the hand, down to 100 Ω if via a large metal object held in the hand.

To simulate a discharge from the human body the circuit shown in Figure 7.7a is used for test purposes:

C_b = capacitance of human body
R_b = resistance of human body
V_b = charging voltage

A primary requirement is that the circuit inductance should be less than 0.1 μH. Different manufacturers may use different values for the above parameters, but the following values are those specified by the IEC[3] and by the US Military Standards[4] and are the ones commonly used:

	IEC	MIL
C_b	150 pF	100 pF
R_b	150 Ω	1500 Ω
V_b	15,000 V	15,000 V
$\frac{1}{2}CV^2$	16.9 mJ	11.3 mJ

Figure 7.7b shows a typical waveform of the discharge current. Its shape is that of an impulse wave with front-time values between 200 ps and 10 ns, and tail-time values between 100 ns and 2 μs. The magnitude of the peak current depends on the discharge voltage V_b and may reach 40A if the discharge voltage is 20 kV. The front time of the

[3] IEC 60801-2 (1991), EMC for Industrial Process Measurement and Control Equipment, Part 2: Electrostatic Discharge Requirements.

[4] MIL-HDBK-263B (1994), *Electrostatic Discharge Control Handbook.*

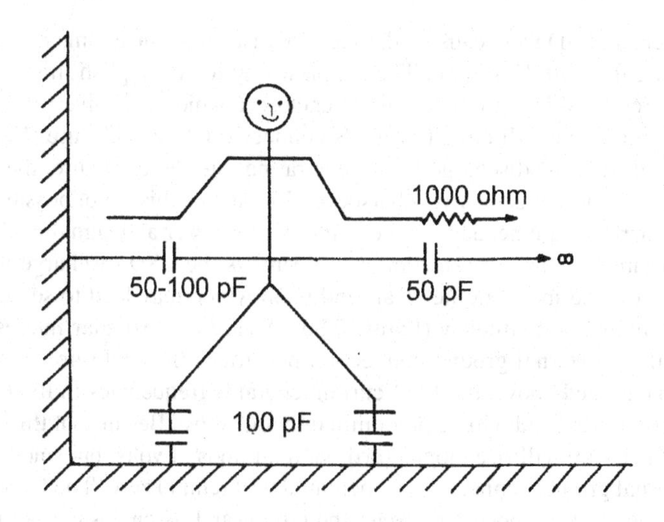

FIGURE 7.6 Self and mutual capacitances of human body.

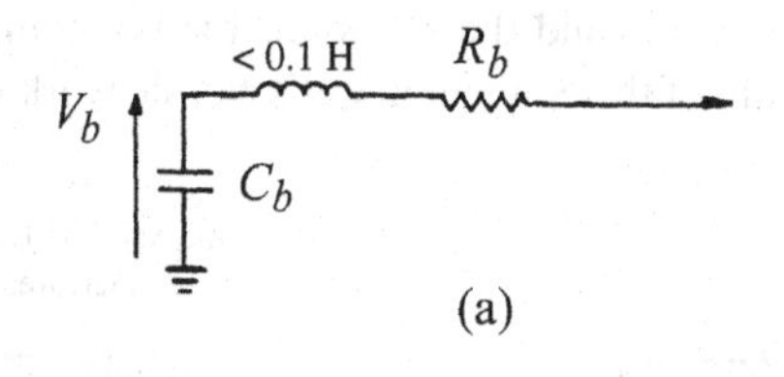

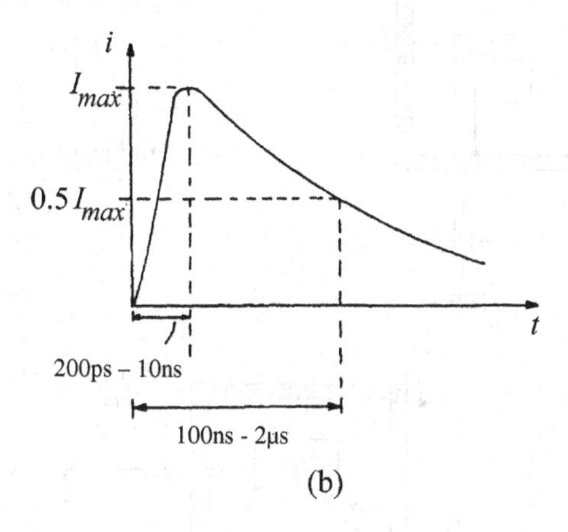

FIGURE 7.7 (a) ESD circuit for human body; (b) typical ESD current waveform.

wave and the discharge energy are the two most important parameters which determine the severity of the discharge. The extremely rapid rise of the wave front current indicates that an ESD contains frequencies in the GHz range so that the inductance of the ground circuit will be of primary importance. Discharges of only a few hundred volts can cause damage to sensitive electronic circuits or cause them to malfunction.

To prevent an ESD from causing damage the discharge current must be prevented from flowing through the circuit. The simplest way to accomplish this would be to place the circuit inside a grounded metal enclosure which will divert the discharge current to ground with all circuit grounds connected to the enclosure (Figure 7.8a). In order to prevent the discharge from penetrating into the enclosure, the enclosure should have no holes or apertures whatsoever. In practice this is not possible because holes and apertures are needed for the entry/exit of power and signal cables as well as for ventilation. In the presence of such openings the ESD discharge will create intense electric and magnetic fields around them which may lead to secondary discharges to the internal circuitry (Figure 7.8b). A case of particular interest is when a cable with an external ground connection penetrates the enclosure (Figure 7.8c). Since, as mentioned above, the ESD current contains frequencies in the GHz range the connection to ground will be predominantly inductive. Because of this the potential of the enclosure will rise momentarily to thousands of volts, but since the circuit has an external ground connection, its potential will remain zero. The large potential difference which thus appears between the circuit and its enclosure can produce a secondary arc which will traverse the circuit. This can be prevented by connecting the circuit to the enclosure. A complete discussion of the corrective measures for the prevention of ESD damage is beyond the scope of this book and interested readers will find an excellent account of this subject in Ott's classic work [17]. Suffice it to say

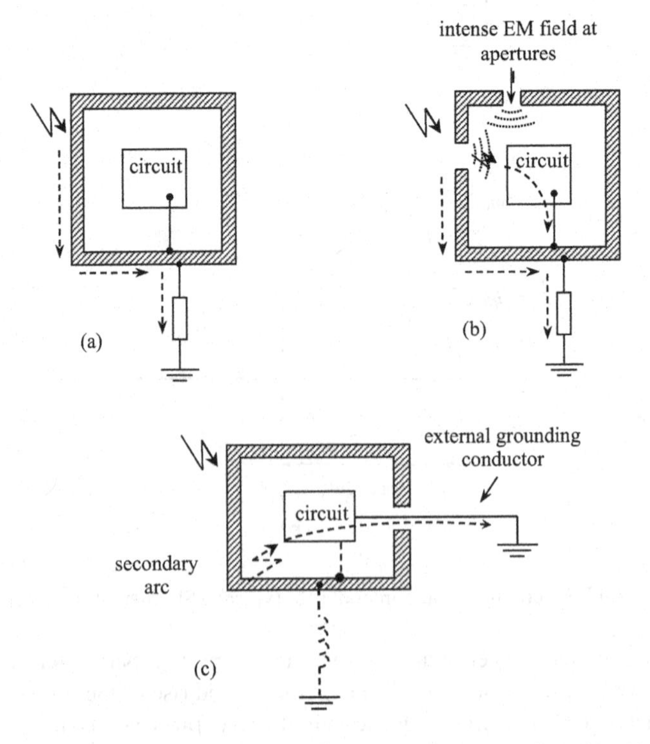

FIGURE 7.8 ESD to enclosure: (a) no openings; (b) with openings;(c) with external ground connection.

here that with a grounded enclosure and good design ESD current can be prevented from flowing through the electronic circuitry.

7.6.2 Brush Discharges and Propagating Brush Discharges

Brush discharges are discharges which occur between charged insulating surfaces and nearby grounded conductors such as equipment or personnel (Figure 7.9). They take the form of filamentary discharge channels (streamers) which look like a brush under low-light conditions (see also corona discharge in the next section). Since the charges on the insulator surface cannot move a single discharge channel is not formed. The maximum energy associated with such discharges is low and does not exceed 4 mJ. However, this may be sufficient to ignite vapors of flammable liquids which have MIEs less than 4 mJ (see Table 7.2).

A potentially more dangerous type of brush discharge is known as a propagating brush discharge. It occurs when a grounded metallic surface has a highly charged insulating layer in the form of a thin film or a deposit of dust or powder (Figure 7.10a).

The combination forms a capacitor[5] that can store a large amount of energy; when the potential difference across the insulating layer exceeds its breakdown strength, it is punctured. The field parallel to the surface then becomes sufficiently great for a massive surface discharge to occur. Such discharges are also known as Lichtenberg discharges and they are potentially very dangerous since their energy is of the order of several joules enough to ignite flammable gases and vapors as well as combustible dusts. Such discharges occur in vessels and pipes which have an insulating lining and in metal surfaces exposed to dusts and powders. Experiment has shown that this type of discharge does not occur for film thicknesses greater than 10 mm or if the breakdown voltage is less than 4 kV for films and less than 6 kV for fabrics.

In some cases a propagating brush discharge can occur over the surface of insulators in the absence of a grounded metal surface (Figure 7.10b). When charges are formed on the inner side of a surface, as for example when a plastic bag is filled with a charged powder or when a liquid flows inside a plastic pipe, these charges may create an external field of sufficient strength either to attract charges of the opposite

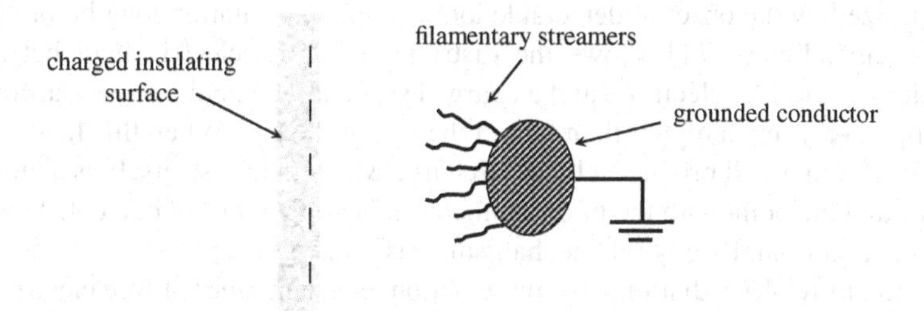

FIGURE 7.9 Brush discharge.

[5] Although a capacitor is usually associated with two conductors, a flat surface charge of uniform density is considered the equivalent of a conducting plane surface.

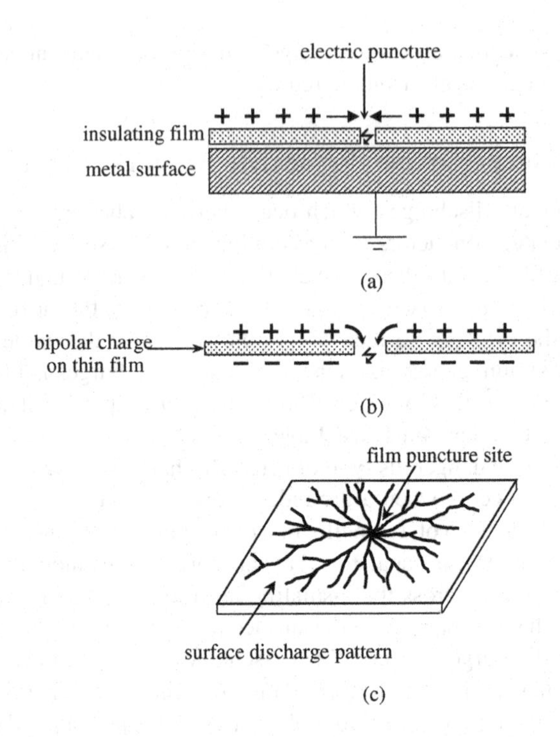

FIGURE 7.10 Propagating brush discharge: (a) insulating film on metal surface; (b) double layer on insulating film; (c) Lichtenberg surface discharge pattern.

polarity or to ionize the air in the immediate vicinity of the surface. In either case a double layer of charge will form across the insulating wall. If the insulation is punctured a propagating brush-type discharge can occur.

7.6.3 Corona Discharges

Corona is a phenomenon associated with highly nonuniform electric fields and is characterized by the onset of detectable ionization and excitation long before breakdown occurs. Figure 7.11 shows the distribution of the electric field between a grounded needle-like electrode and a charged surface. The field intensity around the sharp tip is very much higher than that in the rest of the gap. When this field reaches about 30 kV/cm it will produce a local discharge which manifests itself as a luminous glow around the point with the microammeter indicating a flow of current. This glow is known as corona. Briefly the mechanism is as follows.

Due to ultraviolet radiation, cosmic radiation, etc., a number of free electrons and ions are always present in the atmosphere. When acted upon by an electric field they will be accelerated and after traveling a short distance will collide with air molecules. When the field is sufficiently high the result of such collisions is either the removal of an electron from its molecule (ionization), or the raising of an electron into a higher orbit (excitation). Such orbits are unstable and the electron will drop back to its original orbit and in so doing release the energy originally acquired; this energy appears

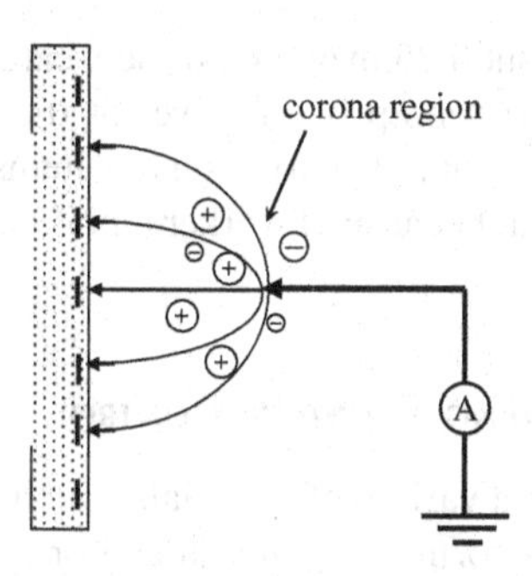

FIGURE 7.11 Electric field between a sharp point and a charged plane.

as visible light or corona. The flow of current is the result of the movement of the generated electrons and ions toward the oppositely charged electrodes. Coronas can be "noisy" both acoustically and electromagnetically.

It should be stressed here that corona is invariably associated with highly nonuniform fields and the energy dissipated is always too low to initiate an explosion. In nonuniform fields where electrodes are not sharp the corona discharge is replaced by a brush discharge which can be considered as a type of high-energy corona discharge.

Since corona is a field effect it can occur either around a live electrode or around a grounded electrode. To distinguish between these two cases it is usual to refer to coronas around live electrodes as *active coronas* and those around grounded electrodes as *passive coronas.*

Besides their use in a large number of industrial applications, corona discharges are used as an effective means of discharging statically electrified surfaces.

7.7 METHODS OF CONTROLLING ELECTROSTATIC CHARGES

7.7.1 Grounding and Bonding

The accumulation of charges on conducting bodies can be effectively controlled by grounding. All parts of an equipment or device should be bonded together and the entire system connected to the supply ground. Grounding allows charges to leek to ground but at the same time if a ground fault occurs during the operation of equipment the potential of any grounded parts will rise to $I_f \times R_g$ above true ground. When all parts are connected together (bonded) and grounded, they will all be at the same potential and there is no danger of a discharge occurring either between them or to ground.

As aforementioned the human body is a good conductor capable of acquiring a charge sufficient to cause a spark discharge which has the energy to ignite flammable material and damage electronic components. As a preventive measure personnel working in the manufacture and assembly of sensitive electronic components and equipment, or other static-sensitive industries, are provided with wrist strap made of conductive material and connected to ground through a 1 MΩ resistor; this resistor will limit the current to about 0.25 mA should the wearer accidentally come in contact with 230 V. Such a wrist strap is effective for the dissipation of static charge acquired by the human body but not charges on any items of clothing. Where the use of wrist straps is not practical because of worker mobility requirements, static control flooring is used.

7.7.2 Floors and Clothing for Static Control

These are floors made from a variety of materials which are made conducting by the addition of carbon in some form (graphite, fibers) or special chemical compounds and include polymers, epoxy based mortars, hard rubbers, etc., either as a continuous layer or in the form of tiles. Their function is to dissipate static charges by grounding all equipment and personnel in direct contact with the floor. They will also significantly reduce the body voltage generated by tribo-electrification. Such floors are used wherever the occurrence of ESDs can have serious consequences such as in the explosives industry, hospital operating rooms, aircraft hangers, manufacture and assembly of sensitive electronic units, or in any areas where there are flammable vapors or gases susceptible to static ignition. There are two types of static control flooring classified according to their resistance to ground: *static conductive floors* and *static dissipative floors.* The distinction between the two types is based on the resistance to the movement of charge across the material's surface. Conducting floors have a much lower resistance than dissipative floors. For the resistance values to be significant they must always be referred to the standard method of measurement used; here we give (Figure 7.12) the most common and practical method as specified by the Standards ASTM F150-06 and ANSI/ESD S7.1-2013.

(a) Point-to-point test (PTP) in which two 2.5″(6.35 cm) diameter electrodes, each weighing 5 lb (2.27 kg) and having a conducting rubber contact surface, are placed 3′ (91 cm) apart at any two points on the floor and the resistance measured by a megger with nominal voltage 10 V dc for conductive floors and 100 V dc for dissipative floors.
(b) Point-to-ground test, also referred to as resistance to ground test (RTG), in which one electrode similar to that described in (a) is placed at any point on the floor surface and is connected to one terminal of the megger. The other megger terminal is connected to an available grounding point. As for the PTP, test megger voltages are 10V and 100 V dc.

In both the above tests all electrodes should be a minimum of 90 cm away from any grounded items or earth ground, and readings taken after 15 seconds. Resistance values are the average of at least five different measurements. If the floor consists of tiles then some of the PTP measurements must be made between electrodes across the seam. Both tests can be carried out on individual floor specimens or on installed floors. For the exact testing procedure the relevant standards should be consulted.

Based on the above measurements the classification of floorings according to their resistance is as follows.

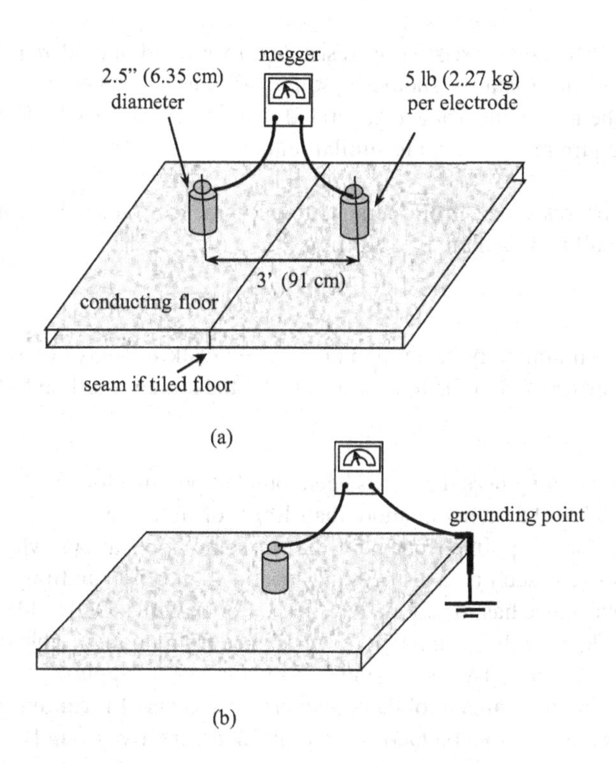

FIGURE 7.12 Measurement of surface resistance: (a) PTP; (b) point-to-ground.

7.7.2.1 Static Conductive Floors

PTP resistance: $> 1 \times 10^4\,\Omega$
Point-to-ground resistance: $< 1 \times 10^6\,\Omega$

These limits have been chosen because if the floor is too conductive personnel may suffer electric shock if exposed to mains voltage, whereas if it is insufficiently conductive charge may not be removed completely or rapidly enough.

7.7.2.2 Static Dissipative Floors

Point-to-ground resistance: $> 1 \times 10^6\,\Omega$ and $< 1 \times 10^9\,\Omega$

Dissipative floors are used wherever rapid rate of charge dissipation can create a magnetic field which could pose problems with the manufacture of electronic components. Otherwise conductive floors are the superior choice.

To be effectively grounded personnel have to wear nonsparking conductive footwear known as ESD footwear. The static control system consists of the person, the ESD footwear, the conducting floor and the connections to ground. The system resistance, which is the resistance from a person's hand to ground, is the sum of the individual resistances $R_{body} + R_{footwear} + R_{floor} + R_{ground}$. According to the Standard ANSI/

ESD S20.20(2014),[6] the wrist strap resistance to ground should not exceed $3.5 \times 10^7\ \Omega$, the system resistance should be $< 1 \times 10^9$, and a walking test is required to ensure that the maximum voltage generated does not exceed 100V. The IEC 61340-5-1(2007)[7] requirements are quite similar: either,

- The total resistance from persons to equipment ground via footwear and floor shall be less than $3.5 \times 10^7\Omega$;

or,

- The maximum body generation (also called walking body voltage) shall be less than 100 V and the total resistance of the system shall be less than $1 \times 10^9\ \Omega$.

It is worth mentioning here that it has been found impossible for a floor with a system resistance < 35 MΩ to generate more than 100 V of static electricity.

It should also be pointed out here that in some applications where hazardous atmospheres exist, such as anesthetizing locations and the handling of explosives, the system resistance has to be less than $10^6\ \Omega$. For example, DOD 4145.26M (2008)[8] specifies that $R_{system} \leq 10^6\ \Omega$ and that the resistance from floor (or table top) to ground must be > 40 kΩ for 110 V supply and >75 kΩ for 220 V supply.

In addition to personnel, mobile equipment, and items of furniture such as chairs and tables should be of conducting material and directly grounded or grounded through conductive rubber castors.

Since the resistance of static control floors and ESD footwear varies with time and with use, their resistances must be periodically measured to ensure that they still comply with their initial specifications and a record should be kept of such measurements.

Figure 7.13 shows the essential features of a typical workstation. They include a static dissipative work surface and ground mat, a personnel grounding wrist strap, a common grounding connection, and appropriate labeling.

In addition to footwear due consideration has to be given to the type of clothing worn by personnel. Electric charges are generated on operators' clothing by tribo-electric effects (rubbing against upholstery, chairs, etc.) but because usual clothing is electrically isolated from the body these charges cannot be dissipated to ground via the skin. To overcome this problem special ESD protective clothing is commercially available. Nonstatic fabric composition varies and is typically as follows:

Polyester 30%–80%
Cotton 30%–75%
Carbon fiber 1%–5%

---

[6] ANSI/ESD S20.20, Development of an ESD Control Program

[7] IEC 61340-5-1, Protection of electronic devices from electrostatic phenomena—General requirements.

[8] DOD 4145.26 M, 2008, Safety Manual for Ammunition and Explosives, C6.4: Static Electricity and Grounding.

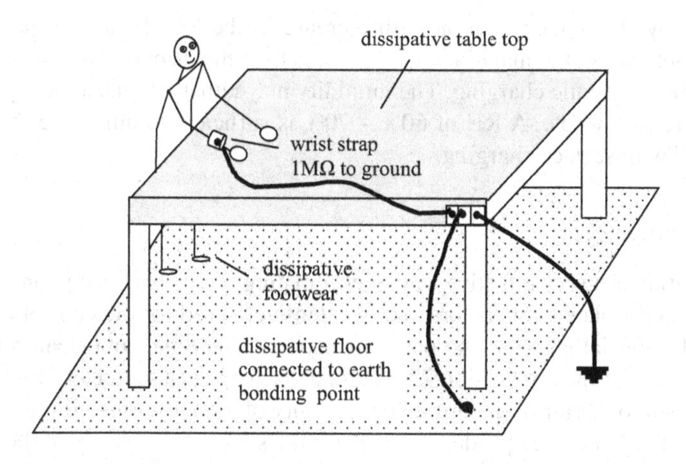

FIGURE 7.13 Schematic diagram of a typical ESD workstation.

Whether special clothes are needed or not depends on the nature of the work environment. In electrostatic protected areas, especially in clean rooms and very dry environments, and where electronic parts being manufactured or handled have a very high degree of sensitivity, ESD protective clothing is required. Such clothing is also necessary in regions where the MIE of flammable vapors is less than 0.2 mJ.

ESD protective fabrics should have a surface resistivity of less than 5×10^{10} ohms. The most quoted standard for the measurement of this resistivity is the European Standard:

EN 1149-1: 2006, Protective clothing—Electrostatic Properties—Part 1: Test methods for measurement of surface resistivity.

The US Standard for protective clothing is,

ANSI/ESD STM 2.1-2018: Electrostatic Discharge Association standard test method for the protection of ESD susceptible items—Garments.

According to this test method the PTP resistance should be as follows:

Garments $R_{p\text{-}p} < 1 \times 10^{12}\,\Omega$
Groundable garments[9] $R_{p\text{-}p} < 1 \times 10^{9}\,\Omega$

7.7.3 Control of Humidity

As mentioned in Section 7.4 the RH of the ambient air has a marked effect on the surface resistivity of materials and hence on the ability of the electrostatically generated surface charges to leak away. The higher the RH the easier it is to dissipate these charges. As shown in Table 7.5 there is a dramatic reduction in body voltage

[9] According to IEC 61340-5-1 "If a groundable garment is used as part of the person's primary ground path (person is connected to a garment which is connected to a grounding cord that is attached to ground) then the maximum resistance from the person's body to ground should be $3.5 \times 10^{7}\Omega$".

generated by tribo-electrification with increase in the RH. If raising the RH does not adversely affect the manufacturing process then this is one of the most effective ways of limiting static charging. The humidity may either be raised locally or in the atmosphere as a whole. A RH of 60% – 70% is sufficient to minimize the dangers associated with surface charging.

7.7.4 Ionization

Air ionization is a very effective way of eliminating static charges on nonconductive materials and isolated conductors and is extensively used for the control of ESD in industrial (especially textile and paper industries), electronic, and clean-room environments. As outlined in Section 7.5 a corona discharge is a source of ions generated by the ionization of air in the immediate vicinity of a highly stressed point (or wire) electrode. The net charge produced has the same sign as the polarity of the point.

In a passive corona discharge the polarity of the point will be opposite to that of the surface charge. Thus for a negatively charged surface positive ions will move away from the positive point toward the surface and neutralize charges there; conversely, if the static charge is positive negative ions will be generated at the point and drawn to the surface. Passive corona will cease once the field at the point is reduced below the value required to initiate corona, and there will therefore only be a partial cancellation of surface charges. Passive ionizers usually consist of a grounded comb-like metal structure consisting of a row of sharp points or teeth on a metal bar or can have a brush-like structure with copper or bronze bristles resembling either a hair brush or a bottle brush.

An active corona source is the best way to reduce static charges to a very low level. The most common type of active ionization device consists of an AC high voltage (2–7 kV) transformer connected to a set of pointed emitters[10] via decoupling resistor and capacitor to ensure that currents drawn if the point is touched are not dangerous. Since the voltage is alternating the corona discharge will generate ions of both signs thus enabling the neutralization of either positively or negatively charged surfaces. However, because such a discharge suffers from a high rate of recombination of positive and negative ions, discharge points should be located no more than 2–10 cm from the charged surface. To reduce the probability of recombination at near distances from the point, filtered compressed air or a fan is used to disperse the ions produced thus increasing the effective range to as much as 70 cm. An active ionizer can neutralize ±1,000 V to ±100 V in less than 10 seconds. It can be switched on and off and can be located wherever most needed. Many ESD workstations are provided with an active ionizer. However, active ionizers must not be used in explosive environments unless specially designed to guarantee that no sparking will occur in the event of a malfunction.

One disadvantage of active ionizers is that electrical discharges in air produce ozone and nitric oxide which, if inhaled for long periods, produce chronic respiratory irritation and therefore require proper venting.

[10] Emitter materials are stainless steel, thoriated tungsten, or single crystals of silicon or germanium, depending upon kind of application. Stainless steel has the highest erosion rate and single crystals the lowest.

Both passive and active ionizers are available commercially in diverse types and sizes.

Some ionizers use a radioactive material, such as radium or polonium, which emits alpha particles to ionize the air. However, the use of such ionizers is very limited as they have to be licensed from the relevant government authorities and strict safety measures have to be enforced. They can be safely used in explosive environments and their size can be extremely small allowing their placement in very confined areas. However, their half-life is relatively short (~130 days) so that they have to be replaced periodically. Their use may also require periodic inspection by the authorities.

8 Protection against Lightning

8.1 NATURE OF LIGHTNING

The need to protect buildings and structures from the damaging effects of lightning gave birth to protective grounding. The phenomenon of lightning itself has been extensively studied and there are a large number of books and hundreds of research papers devoted to it; however, it is relevant here to give the reader a brief outline of the nature of lightning and the lightning stroke to enable him to better understand the protective measures employed.

Lightning is generated by a special type of charged cloud known as a thundercloud. A lightning discharge can occur inside one cloud, between two clouds, or between a cloud and ground. The latter, known as a lightning stroke, is the most dangerous type of discharge against which buildings and structures have to be protected. Many theories have been put forward for the mechanism by which thunderclouds are charged but we will confine ourselves to the results: the lower part of the cloud carries negative charges and the upper part positive charges (Figure 8.1). As the densities of these charges increase so does the electric field intensity associated with them. When this intensity equals the breakdown strength of the ambient air a discharge is initiated either inside the cloud, or between two clouds, or between the cloud and ground.

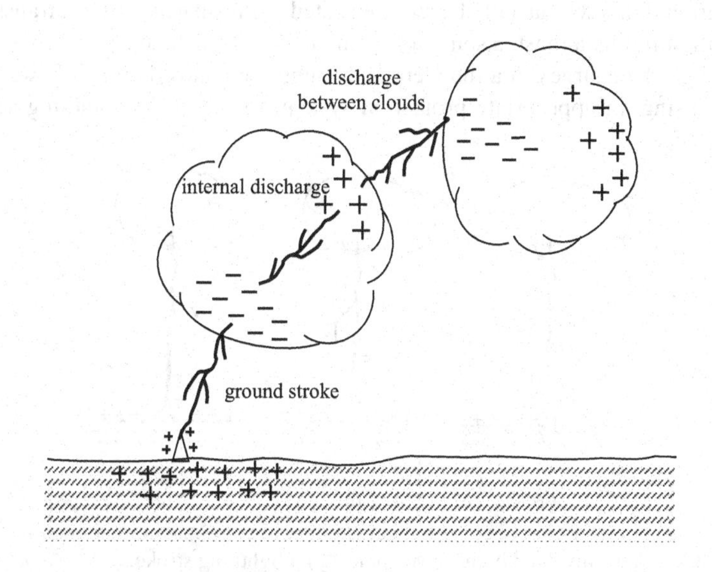

FIGURE 8.1 Types of lightning discharges.

Based on the evidence provided by high-speed photography of lightning strokes we can describe the development of the stroke by the following sequence of ultra rapid events:

1. The electric field between the cloud and ground reaches a value which causes the formation of a negative streamer from the lower part of the cloud directed toward the ground. This pilot streamer (or leader) is faint and difficult to see and since it follows the path of least resistance it appears stepped. As it progresses downward it branches out into several stepped streamers (Figure 8.2a).
2. As the pilot streamer approaches the ground surface a positive upward streamer is initiated from a point on ground where the induced charge density is highest (Figure 8.2b).
3. When the upward and downward propagating streamers meet a large current flows as a highly luminous discharge constituting the main or return stroke (Figure 8.2c). This is followed by a series of subsequent discharges (four strokes on the average) of much lower magnitude than the initial stroke and resulting from the discharge of other portions of the cloud through the conducting path created by the main stroke.

Since the distribution of positive and negative charges inside the cloud is not uniform, the charge density will vary from one part to the other. Similarly the density of induced charge over the surface of the ground will vary according to the conductivity of the soil (mineral ore, wet land, arid sand, etc.). Since a cloud base can cover an area between 5 and 50 km^2, the charge density over such an area will depend both on the nature of the soil and on the ground topography. A lightning stroke will be more likely to occur where the charge density is highest in the cloud and on the ground. On the ground this will usually be some elevated point such as a tree, a transmission line, a tall building, a mast, a spire, a person on an open field, etc.

Although the charges in a thundercloud cannot be controlled, it is however possible, by using an appropriate protection system, to direct the lightning stroke to

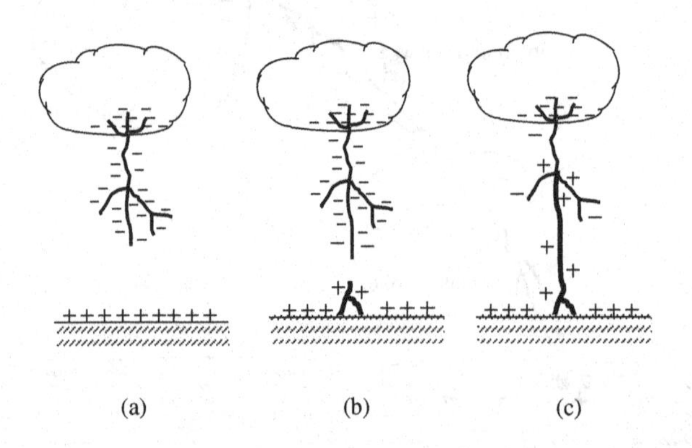

FIGURE 8.2 Steps involved in the development of a lightning stroke.

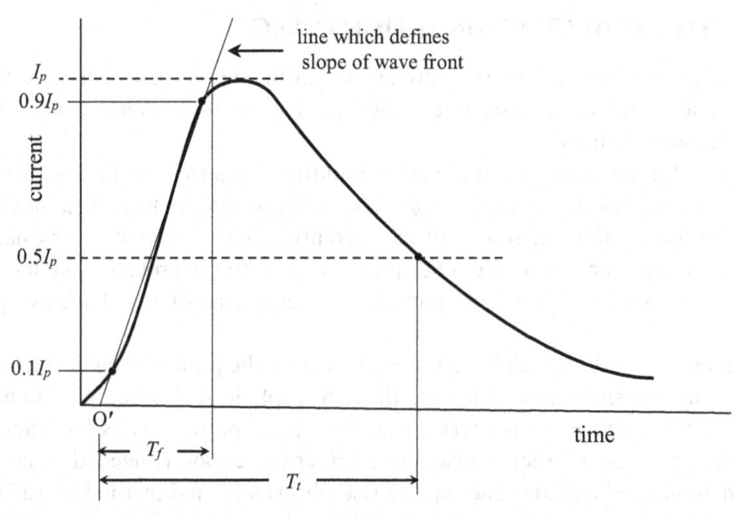

FIGURE 8.3 Waveform and parameters of an initial short lightning discharge current.

specific points on the ground. This is the function of all schemes for the protection of structures against lightning.

Sometimes a positive streamer may start upward, e.g., from mountain peaks or very tall buildings constituting a positive upward leader. However, downward flashes represent the majority of lightning discharges.

Figure 8.3 shows the typical wave shape of an initial lightning stroke. The current rises to a peak value (20–2,000 kA) within an extremely short time (1–20 μs) and then decays with varying rates. Three parameters are needed to identify such waves:

I_p = peak value of current
T_f = time for current to rise to its peak value I_p (front time)
T_t = time for current to drop to 50% of I_p (tail time)

Thus a wave is defined as a T_f/T_t, I_p wave. T_f and T_t are measured from the virtual origin O′ formed by the line joining the $0.1I_p$ and $0.9I_p$ points on the rising part of the curve.

A 10/350 μs wave is used by both the IEC and IEEE standards to characterize the current wave of the initial current stroke and a 8/20μs wave to characterize the current wave of an indirect stroke.[1]

[1] It should be mentioned that the overvoltages created by lightning strikes are characterized by a standard 1.2/50μs voltage wave which is used to test equipments' withstand to overvoltages of atmospheric origin.

8.2 WHY PROTECT AGAINST LIGHTNING

The average number of lightning strokes which hit the earth's surface is 100 per second. The energy of the discharge is dissipated as sound (thunder), heat, light, and electromagnetic radiation.

When lightning strikes an unprotected building or a structure it causes very substantial damage. Metal parts are distorted by the huge electromechanical forces generated (proportional to the square of the current), concrete is shattered by heat, and wooden structures are set on fire. The purpose of a lightning protection system (LPS) is to draw the stroke to itself and provide a direct low-resistance discharge path to ground.

What are the buildings and structures that have to be protected against lightning? It is evident that structures which are liable to explode such as ammunition factories and depots, petroleum refineries, etc., have to be protected. As for other buildings and structures in order to evaluate whether protection is needed or not a risk assessment has to be made. The type of risk philosophy and method of assessment depends on the lightning protection standard to be used. The two most prominent standards are IEC 62305-2010[2] and ANSI/NFPA 780-2017.[3] Although a detailed account of the procedure involved is beyond the scope of the present monograph we have, however, found it pertinent to at least familiarize the reader with the processes involved in risk assessment. In determining the risk factor The IEC Standard[4] considers that the types of loss that a structure and its contents can incur are of prime importance:

- Human hazards, especially the risk of loss of life.
- Loss of production or service to public.
- Damage to structure and/or contents.
- Economic damage: physical damage to equipment and effect on insurance premiums.

Having determined the type of loss a corresponding tolerable risk is determined from tables. This risk is 10^{-5}/year for loss of life or permanent injuries and 10^{-4}/year for other types of losses. A risk of 10^{-5}/year means a one in 100,000 chance of a lightning strike per year. The actual risk is then determined by a series of calculations using formulae and weighting factors for several other aspects. If this actual risk is less than the tolerable risk then no protection is needed.

The IEC standard defines four lightning protection levels (LPLs). For each level a minimum current level to be protected against is designated together with the probability that the current may be greater than these levels:

---

[2] IEC 62305-2010, Protection of Structures against Lightning Pt.1: General Principles; Pt.2: *Risk management*; Pt.3: *Physical damage and life hazards*; Pt.4: *Electrical and Electronic Systems*; Pt.5: *Services*.

[3] NFPA 780-2017, Standard for the Installation of Lightning Protection Systems, Annex L: Lightning Risk Assessment.

[4] This standard has now been adopted by the UK as BS EN 62305 and has replaced BS 6651.

Level	I_{min} (kA)	Probability $I_p > I_{min}$(%)
I	3	99
II	5	97
III	10	91
IV	16	84

I is the highest protection level and IV the lowest.

It is evident that the risk for a given structure will be a function of the frequency of occurrence of lightning strikes at its geographic location and of the size (area) and height of the structure. The higher the frequency and the larger and taller the building, the greater the likelihood of its being struck. The geographical location determines the *number of thunderstorm days per year* that occur in the region, a thunderstorm day being defined as a day in which thunder is heard at least once. This number is known as the *keraunic* (or ceraunic) level T_d of the region. An isokeraunic map of the world is shown in Figure 8.4 based on 1955 records of the World Meteorological Organization (WMO); it is not necessarily 100% accurate today but it provides an example of such maps and a useful guide if local meteorological records are not available.

The keraunic level T_d is an indication of the thunderstorm activity (days per year when thunder can be heard) but does not give an indication of the number of lightning strikes to ground which is of paramount importance when assessing the risk to a structure. It has therefore been replaced by the flash density or lightning strike

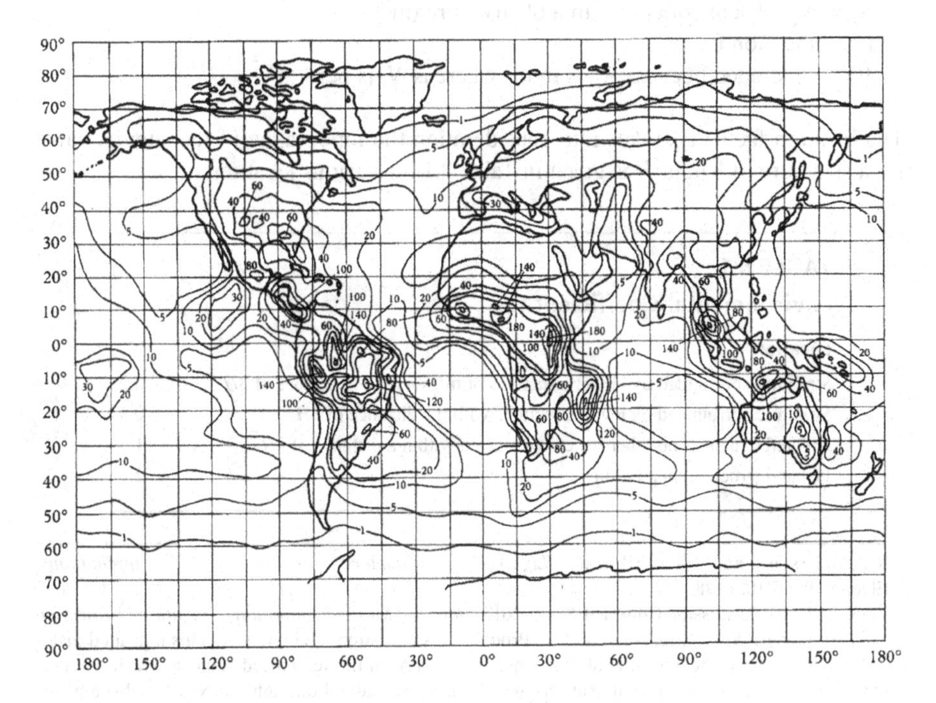

FIGURE 8.4 Isokeraunic map of the world.

frequency N_g defined as the number of flashes to ground per year and per square kilometer. When no information on N_g is available its value can be computed from the relationship,[5]

$$N_g = 0.04T_d^{1.25} \tag{8.1}$$

To get the "feel" of risk assessment we give here the simple lightning risk assessment as given in Annex L of the NFPA standard 780-2017.[6]

In addition to lightning flash density N_g, the other factors which must be taken into consideration are the following:

- Structure location (environment)
- Type of construction
- Structure contents
- Structure occupancy
- Lightning consequences

The expected yearly strike frequency to a particular structure is given by

$$N_D = N_g \times A_D \times C_D \times 10^{-6} \text{strikes/year} \tag{8.2}$$

where

N_g = flash density at structure location
A_D = equivalent collective area of structure (m^2)
C_D = location factor
10^{-6} is included because A_D is in m^2 whereas N_g is per km^2.

The location factor C_D takes into consideration the topography of the site; its value for a structure of height H is given in Table 8.1 for various locations.

TABLE 8.1
Environmental Coefficient C_D

Location	C_D
Structure surrounded by higher structures or trees within a distance of 3*H*	0.25
Structure surrounded by lower structures within a distance of 3*H*	0.5
Isolated structure, no other structures located within a distance of 3*H*	1
Isolated structure on a hilltop	2

[5] R.R.Anderson and A.J.Eriksson, *Lightning parameters for engineering applications*, Electra,69,(65-102),198.

[6] Reprinted with permission from NFPA 780-2017, *Standard for the Installation of Lightning Protection Systems*, Copyright © 2016, National Fire Protection Association, Quincy, MA. This reprinted material is not the complete and official position of the NFPA on the referenced subject, which is represented only by the standard in its entirety which may be obtained through the NFPA website www.nfpa.org.

The effective collective area A_D of a structure is the ground area having the same yearly direct lightning flash probability as the structure. It is obtained by extending a line of slope 1:3 (height of structure to horizontal collection distance) from top of the structure to ground all around the structure as shown in Figures 8.5 and 8.6.

The tolerable lightning frequency N_C is given by

$$N_c = \left(1\times10^{-3}\right)/\,C \text{ events per year} \quad (8.3)$$

where $C = C_2 \times C_3 \times C_4 \times C_5.$

The coefficients take into account the various factors given above and their values are given in Tables 8.2–8.5. The quantity 1×10^{-3} represents the acceptable risk factor for property loss.

To determine if lightning protection is needed the tolerable lightning frequency N_C is compared with the expected strike frequency N_D. If $N_D > N_C$ protection is definitely required; if $N_D \leq N_C$ protection is not needed (optional).

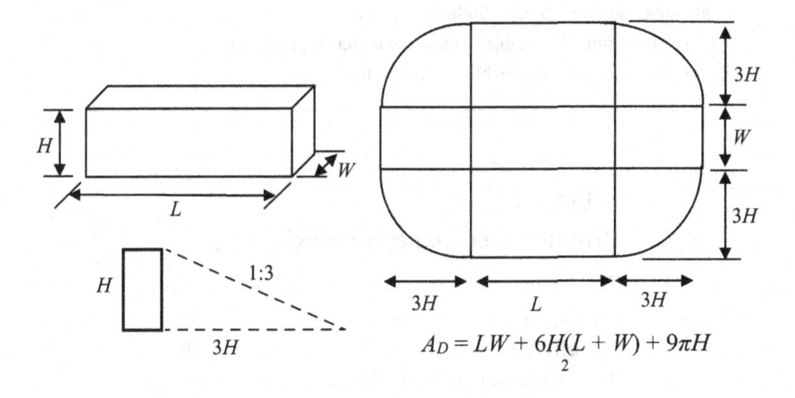

FIGURE 8.5 Equivalent collective area for a rectangular structure.

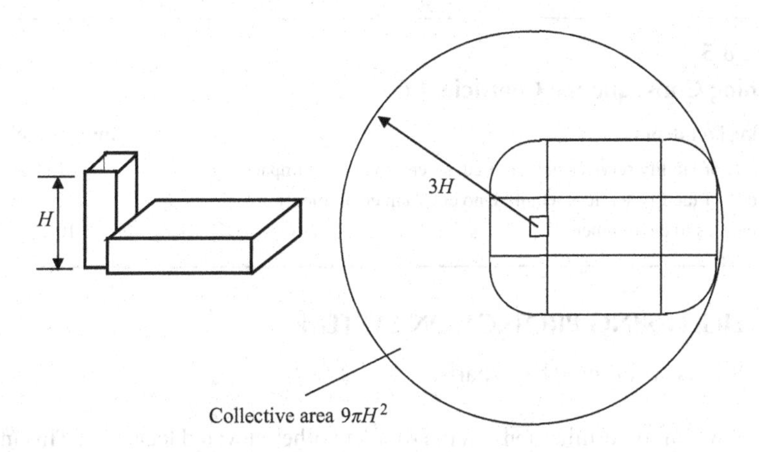

FIGURE 8.6 Equivalent collective area for a rectangular structure with a prominent part which encompasses all portions of the lower part.

TABLE 8.2
Structural coefficient C_2

Structure	Metal Roof	Nonmetallic Roof	Combustible Roof
Metal	0.5	1.0	2.0
Nonmetallic	1.0	1.0	2.5
Combustible	2.0	2.5	3.0

TABLE 8.3
Structure Contents Coefficient C_3

Structure Contents	C_3
Low value and noncombustible	0.5
Standard value and noncombustible	1
High value moderate combustibility	2
Exceptional value, flammable liquids, computer or electronics	3
Exceptional value, irreplaceable cultural items	4

TABLE 8.4
Structure Occupancy Coefficient C_4

Occupancy	C_4
Unoccupied	0.5
Normally occupied	1
Difficult to evacuate or risk of panic	3

TABLE 8.5
Lightning Consequence Coefficient C_5

Lightning Frequency	IndexValue
Continuity of facility services not required, no environmental impact	1
Continuity of facility services required, no environmental impact	5
Consequences to environment	10

8.3 LIGHTNING PROTECTION SYSTEM

An LPS consists of three basic parts:

- A system of terminations on the roof and other elevated locations. This may be either one or more connected vertical rods know as *air terminals*, or a mesh of horizontal conductors known as *air termination networks*.

- A system of down conductors to connect the roof terminations to the grounding system.
- The grounding electrodes.

8.3.1 The Air Terminals

The principle on which air terminals operate is quite simple. As mentioned in 8.1 the main lightning stroke occurs when the negative downward streamer or leader from a cloud meets the positive upward streamer initiated at some point on the ground. The direction of the down streamer is not affected by conditions on the ground until it reaches a certain distance from it which varies between 10 and 100 m. At this distance, which is known as the *striking distance*, the positive streamer will initiate and move upward from some point at which the field intensity is greatest.

The function of the air terminal or lightning rod (Figure 8.7) is to attract the downward leader. The negative charges at the base of the cloud draw the positive charges induced on the ground to the tip of the terminal. These charges create an intense field around the tip sufficient to ionize the surrounding air (this is the same process which gives rise to a corona discharge and discussed in Section 7.5.3). This ionization provides a medium of low resistance which facilitates the development of the positive streamer and its upward propagation from the terminal tip toward the downward streamer and the completion of the main stage of the lightning stroke. Although the electric field at the tip of sharp ends is higher than that at blunt ends, it decreases more rapidly with increasing distance; for this reason rounded tips are today preferred to the pointed ones.

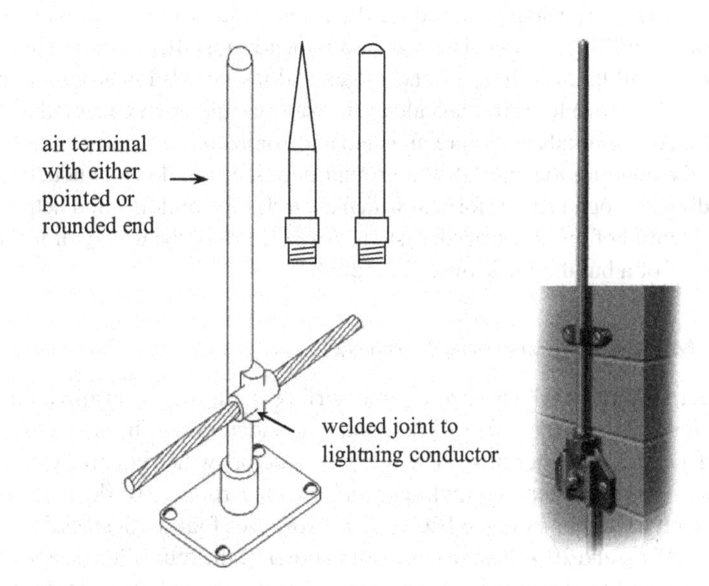

FIGURE 8.7 Examples of an air terminal [Furse].

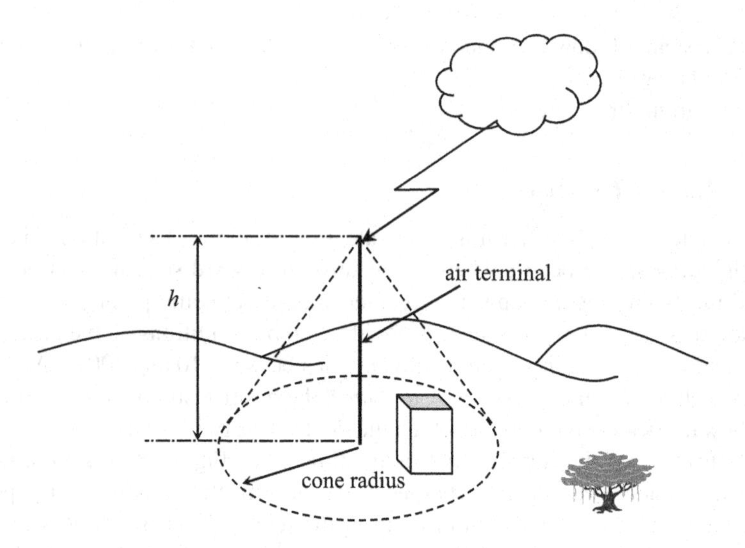

FIGURE 8.8 Cone of protection.

The region protected by an air terminal is known as the cone of protection (Figure 8.8). The majority of standard specifications agree that the protected region is the volume of the cone whose base radius is equal to the terminal height, i.e., a 1:1 cone of protection. All structures which lie inside the cone volume are completely protected against direct strokes. Air terminals in the form of masts are used to protect small isolated buildings, substations, and boats (see also Section 8.5).

Air terminals are usually placed on the roofs of structures to provide a zone of protection around a building. The rods are typically installed around the periphery of flat roofs within 24″(60 cm) of end ridges and (as per NFPA 90) at intervals not exceeding 20 ft (6 m) for terminals along the roof perimeter and 50ft (15m) for inner terminals. All terminals are interconnected and connected to a number of down conductors. By bonding together all the ground points of the down conductors around the building, a ground ring is formed which encircles the building and helps to equalize the potential of the entire earth system. An example of the use of air terminals on the flat roof of a building is shown in Figure 8.9.

8.3.2 Mesh Air Termination Networks

To protect structures which have a large surface area, an air termination network is used. It consists of a number of horizontal conductors which form a mesh on the roof surface and a number of vertical down conductors which connect the horizontal conductors to the system ground. The design of a roof network depends on the specification used. For example IEC 62305-3 specifies four mesh sizes 5 × 5 m, 10 × 10 m, 15 × 15 m, and 20 × 20 m in descending order of the four LPLs (see Section 8.2), with corresponding down conductor spacing 10, 10, 15, and 20 m, respectively. For buildings higher than 60m the mesh network should also cover the vulnerable areas

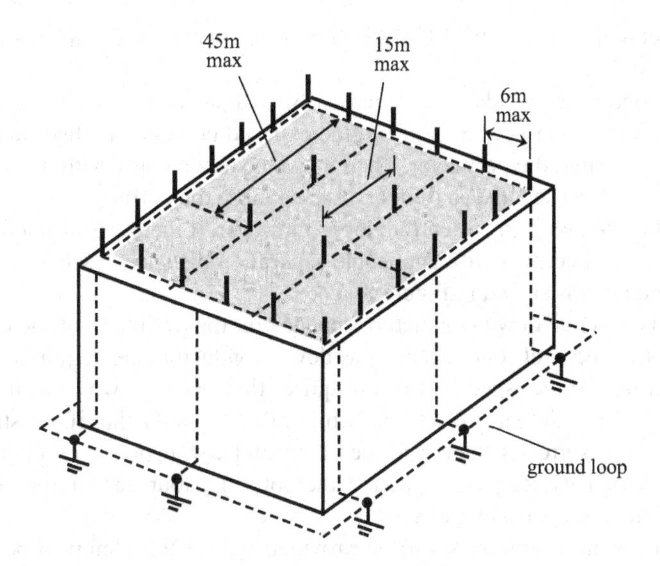

FIGURE 8.9 Air terminals on flat roof.

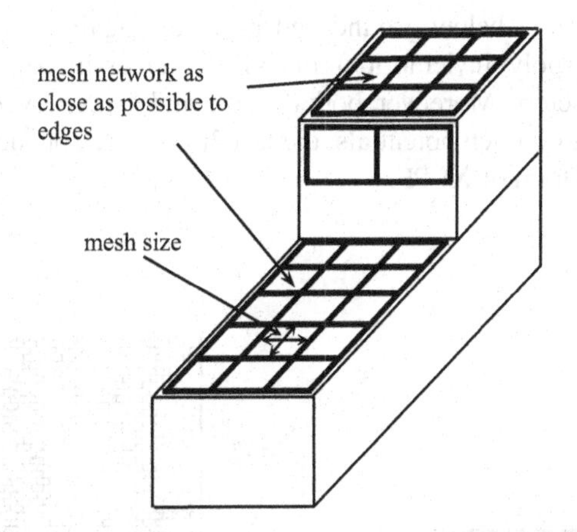

FIGURE 8.10 Example of a mesh air termination network (down conductors not shown).

of the outer walls. With structures which consist of sections having different heights with more than one roof, termination networks of all roofs should be bonded together via their down conductors. Also all metal projections such as masts, aerials, air conditioning units, etc., have to be properly bonded to the air termination network. Figure 8.10 shows an example of an air termination network.

8.3.3 Down Conductors

These conductors provide a low impedance path from the air termination network to the grounding electrode system. When choosing and installing these conductors the following recommendations should be observed:

1. The conductors have to follow the shortest path between the air network and ground.
2. The conductors should be symmetrically mounted on the external surface of the structure walls, starting from the exposed corners and distributed uniformly around the perimeter. Their spacing shall comply with the standard specification used and shall in no case be more than 20 m.
3. It is best to avoid any re-entrant loops; however, if there are such loops then they should comply with applicable separation distance s (see Section 8.4) for the lengths shown in Figure 8.11.
4. The number of down conductors depends on the perimeter of the external edges of the roof, but at least two down conductors are required for any structure. Some specifications require that each down conductor be connected to an earth electrode, while others specify that if the structure perimeter is greater than 76 m then a ground connection is required every 30.5 m of perimeter or a fraction thereof, e.g., four connections will be required for a perimeter of 96 m.
5. Each down conductor has to be provided with a test clamp to be placed 150 cm above the ground surface.
6. The conductors between the test clamps and the ground electrodes should be at least 30 cm below ground and protected against corrosion, e.g., by using PVC (polyvinylchloride) or XLPE (cross-linked polyethylene) - coated conductors. Moreover, both the IEC and the NFPA recommend that, in order to limit touch potentials, the last 3m of bare conductor be insulated with at least 3mm of XLPE.

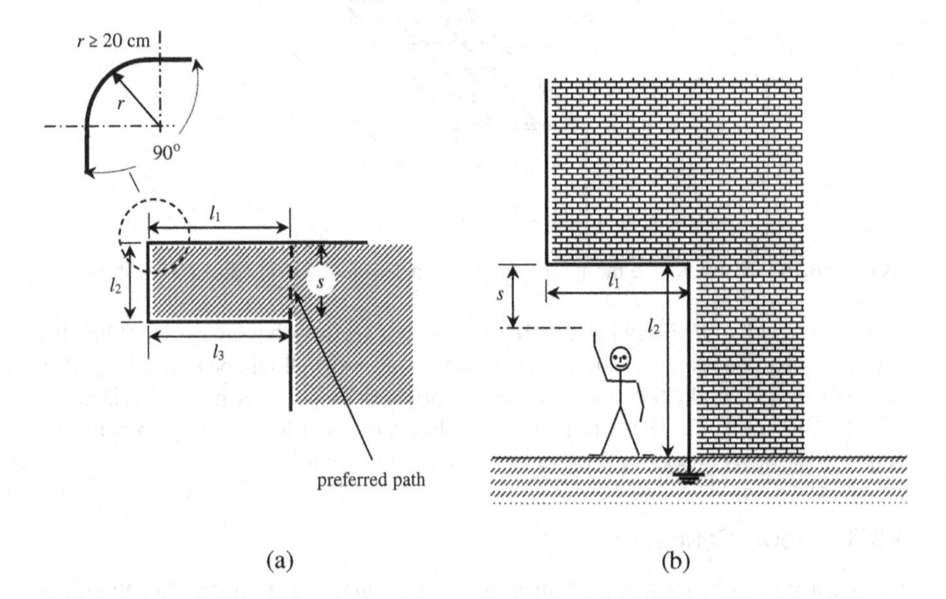

FIGURE 8.11 Down conductor paths and separation distance s for (a) loop and (b) overhang. (a) Acceptable if $l_2 \geq s$, $l = l_1 + l_2 + l_3$; (b) $l = l_1 + l_2$.

7. If reinforced concrete bars of the building are to be used as down conductors they must form a continuous framework within the concrete and equipotential bonding and earthing must comply with IEC 62305-1 or NFPA-780 specifications. Close cooperation between builders and electrical consultants is absolutely necessary from the design stage and throughout the construction stage. If the electrical continuity cannot be guaranteed then separate down conductors must be used.

8.3.4 Types of Conductors

Air termination conductors and down conductors can be copper, aluminum, or galvanized steel. In the case of aluminum or galvanized steel they have to be connected to a copper grounding conductor (aluminum is not allowed) via special bimetallic connectors to prevent corrosion. Moreover aluminum is liable to corrosion when it is in contact with Portland cement and mortar mixes so that special clamps must be used to prevent the conductor from touching the surface. Figure 8.12 shows a selection of different types of clamps used for fixing down conductors to walls and to structure elements.

The normal cross-sectional area[7] of lightning conductors is 50 mm^2. Conductors can be either flat (strip) or solid round or stranded. Until recently the UK favored strip conductors (typically 20 × 2.5 mm), Europe solid round conductors, and the USA

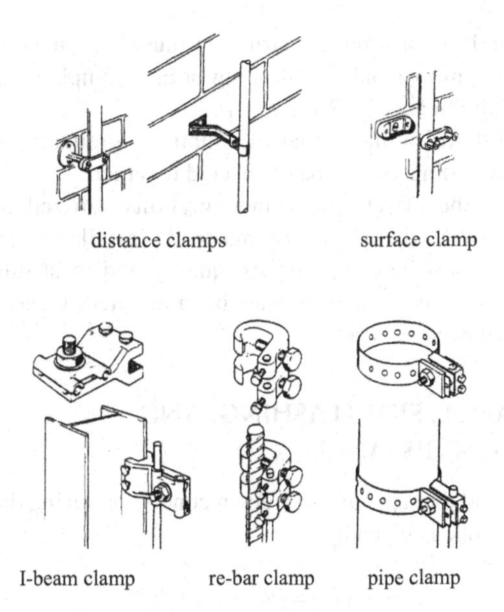

FIGURE 8.12 Fixing clamps for down conductors [Furse].

[7] The section is chosen more for mechanical robustness than for current carrying capacity. Although the current is very large its duration is extremely short. Thus from Eq. (5.1) for a current of 100 kA and duration 50 μs with bolted joints (worst case) the minimum cross-sectional area is 6 mm^2.

stranded conductors. Round conductors have the advantage of being easier to handle and install than flat conductors.

It is possible to use screened insulated cables with a semi conductive outer sheath as down conductors. With these cables there is no risk of side flashing (see Section 8.4) and minimize lightning-induced transients and are thus especially useful if the premises contain sensitive electronic equipment.

8.3.5 Grounding

Proper grounding is a vital part of any lightning protection system. All aspects of grounding have already been considered in the first six chapters of this book. The following additional recommendations should be considered for lightning grounds:

- The resistance to ground should preferably be less than 5 ohms (National Standards have to be consulted for the specified value).
- Exothermic welding of joints is preferred
- The top of ground rods should preferably be at a depth of at least 30 cm below ground using an insulated conductor for the connection between electrode and down conductor. This reduces the voltage gradient around the electrode (see Section 2.1.1).
- It must be assured that the step and touch potentials are within safe limits (see Chapter 4).
- All metal work on or around a structure must be bonded to the lightning protection system to avoid side-flashing or else comply with the separation distance requirements (see Section 8.4).
- Most standards recommend that the lightning grounding system and the protective grounding system be connected together.
- To guarantee the effectiveness and durability (around 30 years) of the protective system, the choice of material, installation procedures, and workmanship must be of the highest quality and must fully comply with the standard specifications that have been adopted. General inspection is recommended every 1–5 years.

8.4 INDUCTANCE, SIDE FLASHING, AND SEPARATION DISTANCE

The voltage to ground that appears on a down conductor during the wave front of the lightning current stroke is given by

$$v(t) = iR + L\, di / dt \tag{8.4}$$

where R is the conductor resistance and L its inductance. For down conductors the inductance is of the order of 1.5 μH/m which is quite small. However, because of the extremely high rate of rise of the current wave front, the inductive component of the voltage becomes predominant. For example, for a front time $T_f = 10\mu s$, a peak current $I_p = 80kA$, and a 25m long conductor,

$$V_L = \frac{25 \times 1.5 \times 10^{-6} \times 80 \times 10^3}{10 \times 10^{-6}} = 300\,\text{kV}$$

This extremely high voltage may lead to an electrical flashover or breakdown between some point on the down conductor and a separately grounded metal body such as a water tank, air-conditioning unit, or any other component of an electrical system. Such flashovers are known as *side flashes* (Figure 8.13). In order not to increase the inductance of a down conductor almost closed loops, such as shown in Figure 8.11, should be avoided if possible.

The rapid rate of rise of current also generates an equally rapidly changing magnetic field which will induce transient voltages and currents in any installation loops and cause damage to sensitive equipment. The induced voltage depends on the mutual inductance M between the down conductor and the loop and is given by

$$V_i = M\frac{di}{dt} \tag{8.5}$$

and as di/dt is almost constant during the wave front, a square voltage will be induced in the loop. The mutual inductance between a down conductor and a rectangular loop (Figure 8.14), which could be a metal window frame, is given by

$$M = \frac{\mu_o b}{2\pi}\ln\frac{s+a}{s} \tag{8.6}$$

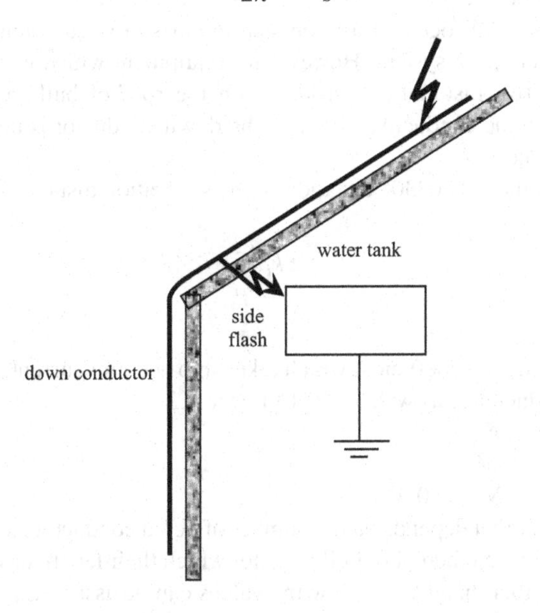

FIGURE 8.13 Side flashing.

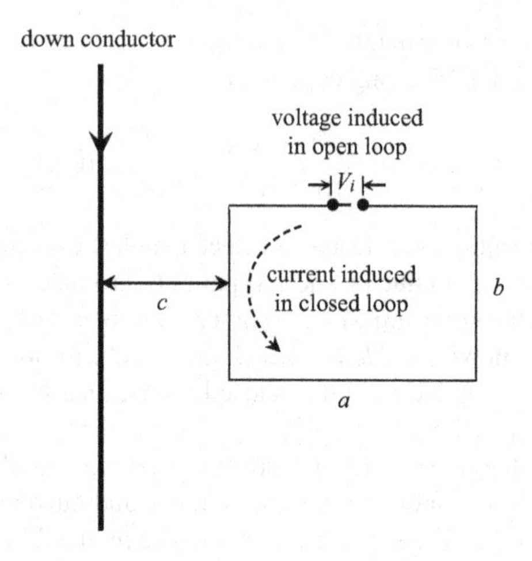

FIGURE 8.14 Current or voltage induced in conducting loops by current in down conductor.

Assuming $s = 1$m, $a = 1$m, $b = 0.5$m, we find that $M = 6.93\mu$H.

For a 10/350μs, 80kA wave, $di/dt = 8\times 10^9$ A/s, so that

$$V_i = 8 \times 10^9 \times 6.93 \times 10^{-6} = 55.4\,\text{kV}$$

If the loop is open this voltage may cause a flashover across the gap. To protect against such lightning-induced voltages sensitive equipment must be provided with surge protection devices.

No side flashes will occur if all conducting parts are equipotentially bonded to the lightning grounding system. However for equipment which is not bonded, such as some conductive installations inside or on the roof of buildings, a *separation distance* between the conductive item and the down conductor is necessary in order to avoid side flashes.

According to the IEC 62305-3 standard, the separation distance is given by

$$s \geq k_i \frac{k_c}{k_m} l \tag{8.7}$$

where

- k_i is an induction coefficient which takes into account current steepness and mutual inductance as well as the LPL (see 8.2):
  LPL I $k_i = 0.08$
  LPL II $k_i = 0.06$
  LPL III and IV $k_i = 0.04$
- k_c is a factor that depends on the number of down conductors and the current sharing between them. For buildings for which their length or width is equal to 4 times their height the following values can be assumed:
  2 conductors $k_c = 0.66$
  3 or more $k_c = 0.44$

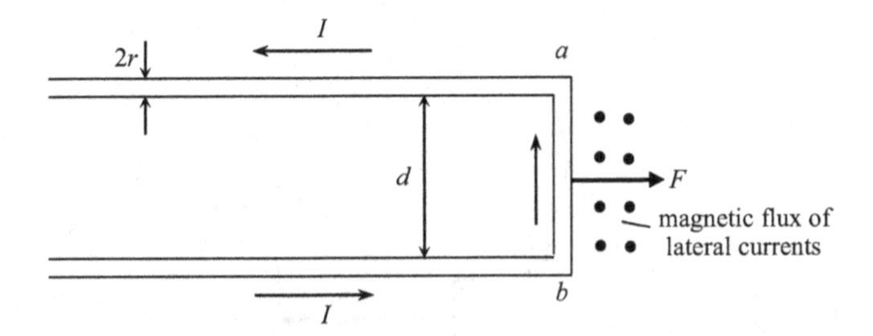

FIGURE 8.15 Electrodynamic force on a bent conductor.

- k_m is a factor which depends on the dielectric strength of the insulation material
  $k_m = 1$ for air
  $k_m = 0.5$ for concrete or bricks
- l is either the total length along the air termination and down conductors from the point where the separation distance is to be considered to the nearest equipotential bonding point or the distance between two points on a down conductor which are closest. For example, in Figure 8.11(a), $l = l_1 + l_2 + l_3$ and in Figure 8.11(b), $l = l_1 + l_2$.
  Assuming a worst case scenario for which

$$k_i = 0.08, k_c = 0.66, k_m = 1$$

then if $l = 25$m, the separation distance required would be $s = 1.32$m.

Another factor to be considered, especially when there are inevitable bends in the down conductors, is the very large electrodynamic force generated between sections of a conductor. This force, repulsive if currents flow in opposite directions and attractive if in the same direction, is proportional to I^2.

The force acting on the section *ab* of the conductor shown in Figure 8.15 and due to the magnetic field of the currents in the lateral parts of the conductor is given by

$$F \cong \frac{\mu_o I^2}{2\pi} \ln \frac{d}{r} \text{ newtons} \tag{8.8}$$

for $d = 10$cm, $r = 0.5$cm, $I = 80$ kA, $F = 3{,}835$ N = 385 kg-force. This very strong force, which tends to straighten the wire, is greatly reduced by replacing the sharp corners with circular bends of radius ≥20cm (8″) as shown in Figure 8.11(a).

8.5 ZONES OF PROTECTION

The most widely used method for determining the areas of a structure or group of structures that need protection and hence most appropriate locations of air terminals or air termination network is the so called Rolling Sphere Method (RSM). The method is based on the concept of the striking distance. As mentioned in Section 8.3.1 this is the distance which a positive streamer, initiated at some point on a structure on

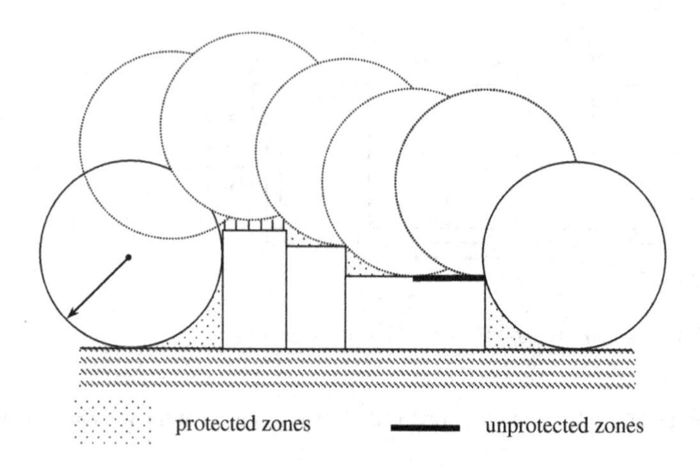

FIGURE 8.16 Path of rolling sphere for identifying surfaces which need protection.

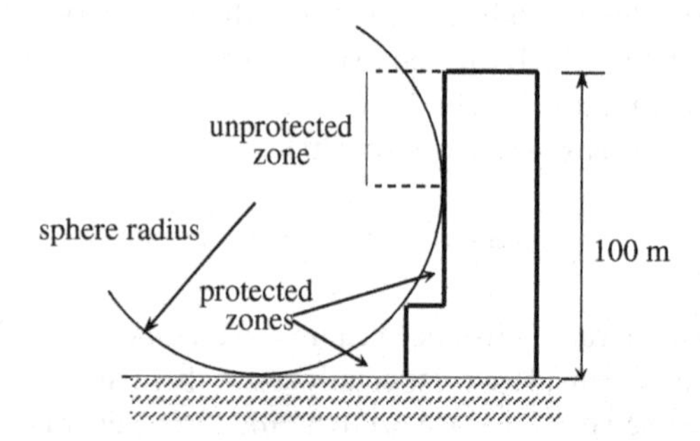

FIGURE 8.17 Protected and unprotected zones for a tall building.

the ground, travels upward to join the descending streamer, and produce the return stroke. The initiation and propagation of the positive streamer from some point depend on the field intensity between that point and the negative charges at the tip of the downward leader reaching a critical value. The more the number of charges at the tip, the greater the distance at which the critical field is attained and hence the greater the striking distance and the higher the discharge current.

The tip of the downward leader is considered to be located at the center of an imaginary sphere whose radius is equal to the striking distance. This sphere is then rolled over the building (Figure 8.16) and any part that can be touched by the sphere will be susceptible to a lightning strike. Those parts of the building that cannot be touched are considered to lie within a zone of protection.

When there is a risk that a lightning strike to the side of a tall structure may cause damage (Figure 8.17) then an extension of the air termination network to those parts should be considered.

American standard NFPA 780 specifies a rolling sphere radius of 150 ft (45.7 m) for normal use but this radius is reduced to 100 ft (30 m) for structures containing flammable liquids and gases. The new IEC Standard 62305-1 has four different sphere radii corresponding to four LPLs mentioned in Section 8.2. These are

Level	I_{min} (kA)	Sphere Radius (m)
I	3	20
II	5	30
III	10	45
IV	16	60

For each level the minimum peak current value (kA) is used to determine the sphere radius S using the formula

$$S = 10I^{0.65}(\text{m}) \tag{8.9}$$

The RSM may also be used to determine the protection zones in a substation. The sphere radius is determined from the following formula adopted by the IEEE[8]

$$S = 8k\,I^{0.65}(\text{m}) \tag{8.10}$$

k = 1.0 for strokes to wire and ground plane,
= 1.2 for lightning strokes to masts,
I = return stroke current in kA.

Figure 8.18 shows an example of such an application. From the maximum height of the substation equipment, the size of the substation and a rolling sphere radius, the height and number of masts, and their distance apart can be determined. Mast heights range from 3 to 15m and their number from 1 to 4. For further details the reader is referred to the IEEE standard.

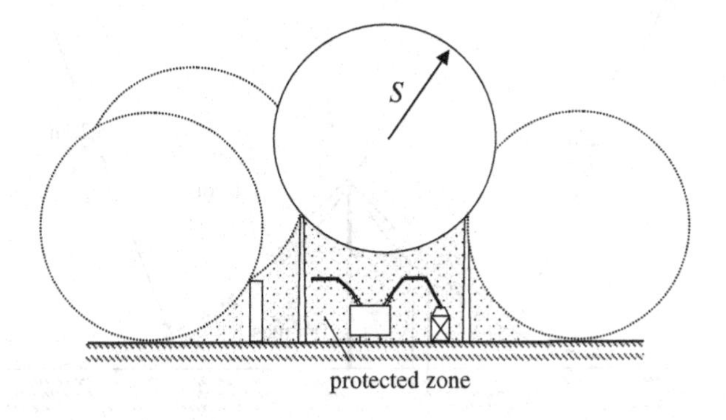

FIGURE 8.18 RSM for identifying zones of protection in a substation with protective masts.

[8] IEEE Std 998-2017, Guide for Direct Lightning Stroke Shielding of Substations.

In the 1:1 cone mentioned in Section 8.3.1 the protective angle is 45°. Such a cone can be safely applied to determine the zones of protection for small substations and simple-shaped structures of height up to 10m for level I protection and 30m for level IV protection. In Figure 8.19 a 45° triangle is superposed on a scaled profile drawing of the structure such that the cone apex coincides with the tip of the air terminal whose protective zone is to be determined. Figure 8.20 shows a diagram drawn to scale for comparing the zones of protection provided by a 10m high mast using a 1:1 cone and a 20m radius rolling sphere corresponding to the highest protection level; the two zones are almost identical. For taller structures there are significant differences between the two methods and cone protective angles as small as 23° are necessary. It should be mentioned here that the standard IEC 62305-1 includes a detailed and somewhat laborious protection angle method in which, for each LPL, the cone angle is given as a function of the height of the air rod above the ground plane for each protection level.

In general the RSM is the preferred method used for identifying the surfaces of structure that need protection because it simpler, safer, and can be used for all types of structure and for all levels of protection required.

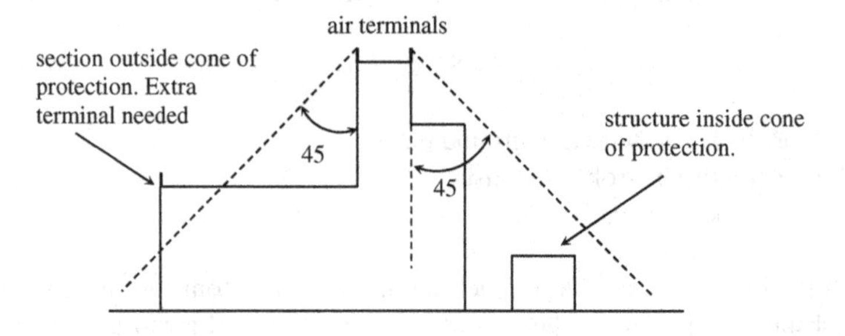

FIGURE 8.19 Cone of protection of a structure (profile must be drawn to scale).

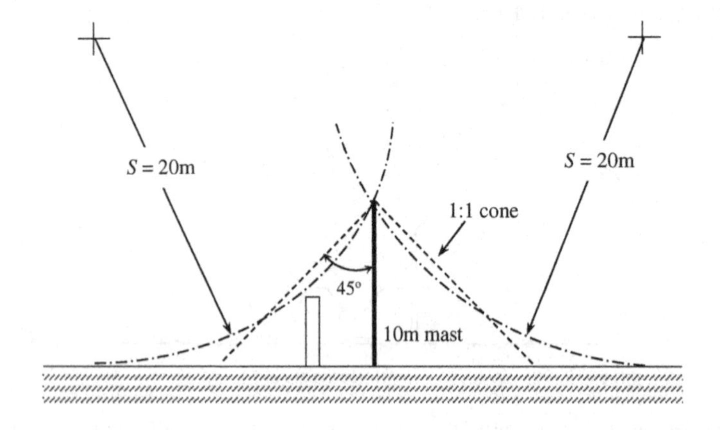

FIGURE 8.20 Comparison between protected zone of a 1:1 cone of protection and rolling sphere for a 10m mast (diagram drawn to scale).

8.6 TANKS AND STACKS

Steel tanks with fixed or floating metal roofs and containing flammable liquids or gases will not need protection against lightning provided that (a) all metal parts are at least 3/16″ (4.76 mm) thick and (b) their base is in intimate contact with ground all around their perimeter to a depth of ≥50 cm. It must be ensured that the resistance to ground is less than 5 ohms or has the value specified by the standard adopted.

Tanks with nonmetallic roofs have to be protected with air terminals, masts, or aerial ground wires as dictated by experience and as required either by NFPA 780 or by any other relevant National Standard.

All joints, couplings, and pipe connections have to be electrically continuous and all vapor or gas openings closed or flame proof.

Stacks (chimneys) for flue gases and dusts must be protected by air terminals at the top of the stack and uniformly distributed around the perimeter with a maximum separation of 2.5 m. The length of the terminals depends on the type of gas or dust emitted. It ranges from 0.5 m for nonflammable emissions to 1.5 m in the case of explosive gases or dusts. If forced draught is used then the length of the terminals should not be less than 4.5 m. Moreover, if the flue gases are corrosive a 1.6 mm thick lead coating on the air terminals is required. In all cases a minimum of two down conductors are required.

The above information should be used as a preliminary guide. For a comprehensive protection policy the complete designated standard which is to be used must be considered in its entirety.

8.7 PROTECTION OF TRANSMISSION LINES BY AERIAL GROUND WIRES

Because overhead power transmission lines supported by tall towers run for long distances through mostly flat territory, they are perhaps one of the structures that are most vulnerable to direct lightning strokes. Overhead ground conductors can provide a degree of protection against such strokes. One or more ground conductor is placed on top of power transmission towers (Figure 8.21) and extends over the entire length of the line. This conductor acts as a shield which protects the line conductors and towers from atmospheric electricity and especially from direct lightning strokes. In this section we give the theoretical background to clarify this.

Measurements have confirmed[9] that under fair weather conditions there is a constant electric field which extends vertically downward from the space charges present in the upper atmosphere to earth. If the strength of this field is E_0 then the absolute potential (assuming that the potential at the earth's surface is zero) at a point P at a height z above the earth's surface (Figure 8.22) is

$$V_o = E_o z \tag{8.11}$$

[9] B.F.J.Schonland, *Atmospheric Electricity*, Methuen, London, 1953.

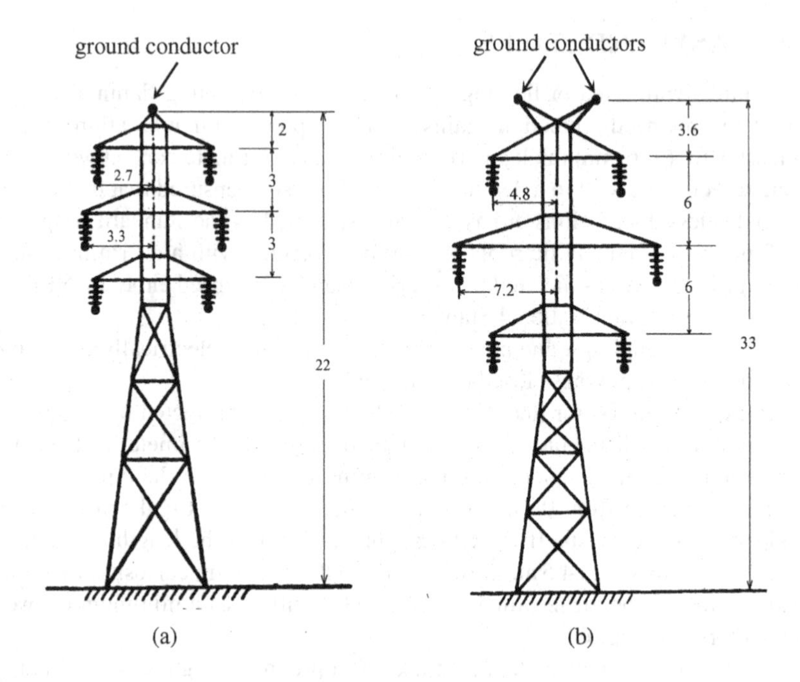

FIGURE 8.21 Ground conductors on typical towers: (a) 66kV; (b) 220kV (dimensions in m).

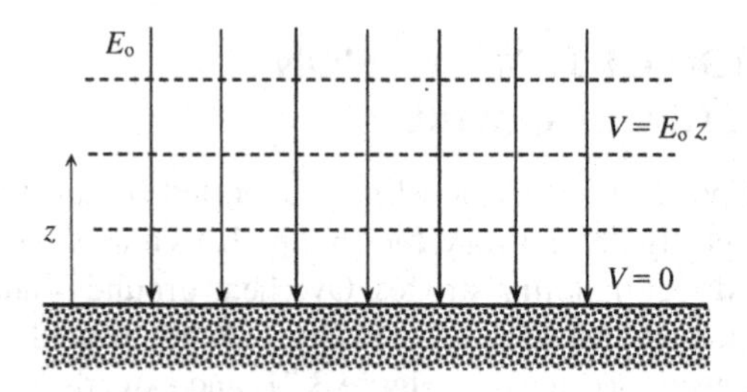

FIGURE 8.22 Fair weather atmospheric field.

and the equipotential surfaces in this case are horizontal planes. If an isolated conductor is placed at any point in space it will acquire the potential of this point but no electric charge. When the conductor is connected to ground it will acquire an induced charge (see Section 5.7) and the secondary field produced by these charges will modify the primary uniform as follows.

8.7.1 Secondary Field of Ground Conductor

Assume that the charge per unit length on the ground conductor is λ C/m and that the conductor is at a height h above ground. For the potential to be zero at all points on the ground plane it is necessary to assume the presence of an image conductor

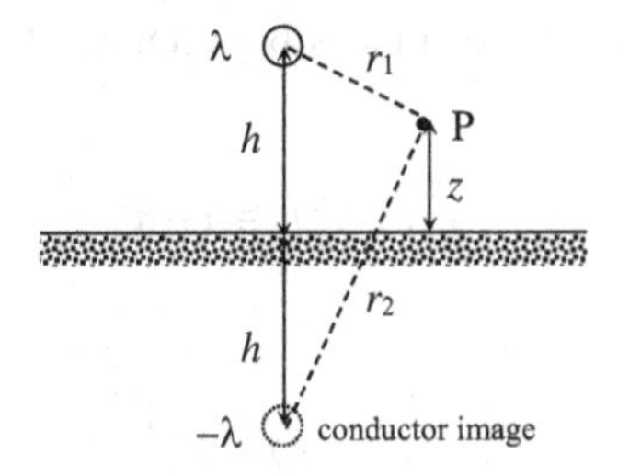

FIGURE 8.23 To determine potential at P due to a line charge and its image.

at an equal distance h below ground with an equal and opposite charge $-\lambda$ C/m. The potential at any point P (Figure 8.23) due to these two line charges is

$$V_{Pg} = \frac{\lambda}{2\pi\varepsilon_o}\ln\frac{r_2}{r_1} \tag{8.12}$$

At the ground surface where $z = 0$, $r_1 = r_2$, the potential is zero according to Eqs. (8.11) and (8.12); it must also be zero at the surface of the ground conductor, $r_1 = r$, $r_2 \sim 2h$, $z \sim h$ ($h >> r$). We thus have that

$$(V_o + V_{Pg}) = E_o h + \frac{\lambda}{2\pi\varepsilon_o}\ln\frac{2h}{r} = 0$$

$$\frac{\lambda}{2\pi\varepsilon_o} = -\frac{E_o h}{\ln\frac{2h}{r}}$$

It is evident that the charge which appears on the conductor is negative and by substituting this value in Eq. (8.12) we obtain

$$V_{Pg} = -E_o h\frac{\ln(r_2 / r_1)}{\ln(2h / r)} \tag{8.13}$$

The total potential at P is thus

$$V_P = V_{Po} + V_{Pg} = E_o z - E_o h\frac{\ln(r_2 / r_1)}{\ln(2h / r)} \tag{8.14}$$

The relative change in the potential at the point P due to the presence of the ground wire is

$$\eta = \frac{V_{Po} - V_p}{V_{Po}} = -\frac{V_{Pg}}{V_{Po}} = \frac{h}{z}\frac{\ln(r_2 / r_1)}{\ln(2h/ r)} \tag{8.15}$$

If $z = h'$ and assuming that the distances between conductors are small compared with h we may write

$$r_2 \cong h + h'; r_1 \cong h - h'$$

and Eq.(8.15) becomes

$$\eta = \frac{h}{h'} \frac{\ln\left[\frac{h+h'}{r_1}\right]}{\ln\frac{2h}{r}} \tag{8.16}$$

The ratio η indicates the effectiveness of the ground conductor in shielding the power conductors from the atmospheric field. For example, if we assume the following practical values, $r_1 = 3\,\text{m}$, $h' = 19\,\text{m}$, $h = 22\,\text{m}$, $r = 6\,\text{mm}$.

We find that

$$\eta = \frac{22}{19} \frac{\ln(41/3)}{\ln(44/0.006)} = 34\%$$

It should be pointed out here that the effect of the diameter of the ground conductor and of its distance from a power conductor is limited since these distances appear in the logarithmic term.

The electric field intensity in the direction of r_1 is from Eq.(8.14)

$$E_{Pr_1} = -\frac{\partial V_P}{\partial r_1} = -\frac{h}{r_1} \frac{E_o}{\ln(2h/r)}$$

and on the surface of the conductor ($r_1 = r$)

$$E_g = -\frac{h}{r} \frac{E_o}{\ln(2h/r)} \tag{8.17}$$

The negative sign indicates that the direction of the field is toward the conductor surface; using the above values ($h = 22\,\text{m}$, $r = 6\,\text{mm}$) we find from the above equation that

$$|E_g| = \frac{3{,}667}{8.9} E_o = 412 E_o$$

The average value of the atmospheric field E_o under fair weather conditions is 100 V/m and in the presence of thunderclouds it may rise to between 10 and 300 kV/m. If we assume that E_o = 10 kV/m we find that

$$|E_g| = 4{,}120\,\text{kV/m} = 41.2\,\text{kV/cm}$$

This field is sufficient to produce a corona discharge around the conductor and draw lightning to it thus protecting the power line.

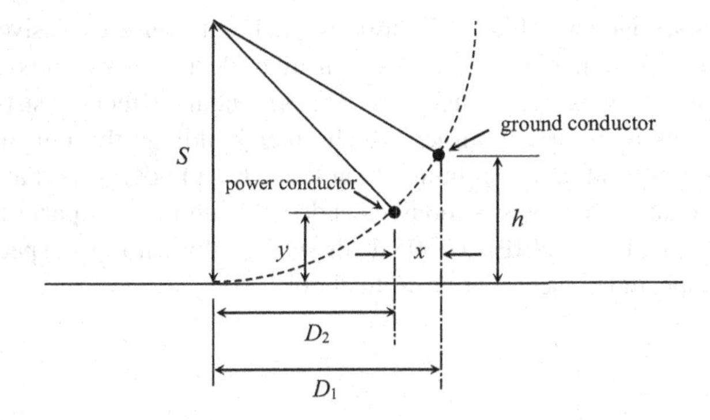

FIGURE 8.24 Ground conductor height and striking distance.

8.7.2 Height of Ground Conductor

To determine the minimum height of the ground conductor we shall make use of the RSM in which the radius of the sphere represents the striking distance S as described in Section 8.5. If h is the height of the ground wire, y is the height of the highest line conductor, and x is the distance between them, then from Figure 8.24 we have that

$$D_1^2 = S^2 - (S-h)^2$$

$$D_1 = \sqrt{h(2S-h)}$$

Similarly,

$$D_2 = \sqrt{y(2S-y)}$$

so that

$$x = D_1 - D_2 = \sqrt{h(2S-h)} - \sqrt{y(2S-y)} \tag{8.18}$$

For given values of S, y, and x, the minimum value of h can be determined from Eq. (8.18). The striking distance can be determined from Eq. (8.10). As an example, for $I = 50$ kA we find that

$$S = 8 \times 50^{0.65} = 100\,\text{m}$$

and for $y = 20\,\text{m}$, $x = 4\,\text{m}$, $S = 100\,\text{m}$, we find from Eq.(8.18) that $h = 23\,\text{m}$.

8.8 PROTECTION AGAINST SURGES

In this chapter we have been mainly concerned with the protection of structures against direct lightning strikes, a topic intimately associated with grounding which is the main topic of this book. However, we must draw the attention of the reader that there is a secondary effect of lightning which can cause extensive damage to all types of equipment, particularly the sensitive electronic systems and devices which permeate every aspect of life today. This secondary effect consists of induced transient voltages and current surges which enter buildings through mains power supplies, telephone and data communication lines, both underground and overhead. Protection against such surges is today considered as an integral part of a lightning protection scheme. Part 4 of IEC 62305 deals specifically with this aspect which lies beyond the scope and subject matter of this book.

Appendix A: Wire Sizes

TABLE A.1
Cross-Sectional Area and Diameter of Copper Conductors (AWG)

Size		Stranding		Overall	
AWG/ (kcmil)	Area (cmil)	Quan-tity	Diam. (in.)	Diam. (in.)	Area (in.^{2})
18	1620	1	–	0.040	0.001
18	1620	7	0.015	0.046	0.002
16	2580	1	–	0.051	0.002
16	2580	7	0.019	0.058	0.003
14	4110	1	–	0.064	0.003
14	4110	7	0.024	0.073	0.004
12	6530	1	–	0.081	0.005
12	6530	7	0.030	0.092	0.006
10	10380	1	–	0.102	0.008
10	10360	7	0.038	0.116	0.011
8	16510	1	–	0.126	0.013
8	16510	7	0.049	0.146	0.017
6	26240	7	0.061	0.164	0.027
4	41740	7	0.077	0.232	0.042
3	52620	7	0.087	0.260	0.053
2	66360	7	0.097	0.292	0.067
1	83690	19	0.066	0.332	0.087
1/ 0	105600	19	0.074	0.373	0.109
2/0	133100	19	0.084	0.419	0.138
3/0	167800	19	0.094	0.470	0.173
4/0	211600	19	0.106	0.528	0.219
250	–	37	0.082	0.575	0.260
300	–	37	0.090	0.630	0.312
350	–	37	0.097	0.661	0.364
400	–	37	0.104	0.728	0.416
500	–	37	0.116	0.813	0.519
600	–	61	0.099	0.893	0.626
700	–	61	0.107	0.964	0.730
750	–	61	0.111	0.998	0.782
800	–	61	0.114	1.03	0.834
900	–	61	0.122	1.094	0.940
1000	–	61	0.128	1.152	1.042
1250	–	91	0.117	1.289	1.305
1500	–	91	0.128	1.412	1.566
1750	–	127	0.117	1.526	1.829
2000	–	127	0.126	1.632	2.092

TABLE A.2
Cross-Sectional Area of Copper Conductors According to German,[a] British,[b] and USA[c] Standards

Nominal Cross-Sectional Area (mm²)	Nominal Diameter (mm)	Conductor Size Designation
0.65	1.1	NBS 3/.020
0.65	0.92	NBS 1/.336
0.75	1.02	0.75
0.821	1.12	AWG 18
0.97	1.2	NBS 1/.044
1	1.15	1
1.039	1.58	AWG 17
1.29	1.29	NBS 3/.029
1.307	1.4	AWG 16
1.5	1.65	1.5
1.652	1.99	AWG 15
1.94	1.63	NBS 3/.036
1.94	1.85	NBS 1/.064
2.083	1.8	AWG 14
2.5	2.08	2.5
2.625	2.21	AWG 13
2.9	2.34	NBS 7/.029
3.309	2.3	AWG 12
4	2.61	4
4.17	2.75	AWG 11
4.52	2.95	NBS 7/.036
5.26	2.8	AWG 10
6	3.3	6
6.45	3.71	NBS 7/.044
6.634	3.97	AWG 9
8.366	3.6	AWG 8
9.35	4.17	NBS 7/.052
10	4.67	10
10.55	4.88	AWG 7
13.296	5.2	AWG 6
14.52	5.23	NBS 7/.064
16	5.59	16
16.767	5.89	AWG 5
19.35	6.5	NBS 19/.044
21.15		AWG 4
25		25

(*Continued*)

TABLE A.2 (*Continued*)
Cross-Sectional Area of Copper Conductors According to German,[a] British,[b] and USA[c] Standards

Nominal Cross-Sectional Area (mm^2)	Nominal Diameter (mm)	Conductor Size Designation
25.81	6.61	NBS 19/.052
26.662	6.6	AWG 3
33.625	7.42	AWG 2
35	7.8	35
38.71	8.13	NBS 19/.064
42.406	8.43	AWG 1
48.5	9.3	NBS 19/.072
50	9.47	50
53.508	10.6	AWG 1/10
64.52	10.62	NBS 19/.083
67.442	11	AWG 2/0
70	11.94	70
77.5	12.8	NBS 37/.064
85.024	12.8	AWG 3/0
95	13.41	95
96.77	14.5	NBS 37/.072
107.218	14.61	AWG 4/0
120	14.8	120
126.675	16.2	MCM 250
129.03	16	NBS 37/.083
150	17.3	150
152.01	18	MCM 300
162	18.4	NBS 37/.093
177.345	18.49	MCM 350
185	19.56	185
193.55	20.5	NBS 37/.103
202.68	20.65	MCM 400
228.02	21.3	MCM 450
240	21.72	240
253.35	22.68	MCM 500
258.06	23.6	NBS 61/.093
278.71	23.6	MCM 550
300	24.49	300
304	25.35	MCM 600
322.58		NBS 61/.103
329.35		MCM 650
354.71		MCM 700
380		MCM 750
400		400

[a] VEB.
[b] NBS (New British Standard).
[c] AWG& MC.

TABLE A.3
Resistance of Copper Conductors

Cross-Sectional Area (mm^2)	Resistance (70°C) (Ω/km)
0.75	29
1	21.7
1.5	14.7
2.5	8.71
4	5.45
6	3.62
10	2.16
16	1.36
25	0.863
35	0.627
50	0.463
70	0,321
95	0.232
120	0.184
150	0.15
185	0.1202
240	0.0922
300	0.0745

Bibliography

[1] G.T. Tagg, *Earth Resistances*, Newnes, London, 1964.

[2] V.Manoilov, *Fundamentals of Electrical Safety*, MIR Publishers, Moscow, 1975.

[3] W.F.Cooper, *Electrical Safety Engineering*, Butterworth, Oxford, 1994.

[4] *Recommended Practice for Grounding of Industrial and Commercial Power Systems (Green Book)*, IEEE Std. 142, 2007.

[5] *The National Electrical Code*, NFPA 70, 2017.

[6] *Earthing*, British Standard Code of Practice, CP 1013, 1965.

[7] *Guide for Safety in AC Substation Grounding*, IEEE Std 80-2000, (ANSI).

[8] *Military Handbook*, MIL – HDBK – 419A, 1987.

[9] A.P. Sakis Meliopoulos, *Power Grounding and Transients*, Marcel Dekker, New York, 1988.

[10] G. Luttgens and N. Wilson, *Electrostatic Hazards*, Butterworth, Oxford, 1997.

[11] L.B. Loeb, *Static Electrification*, Springer-Verlag, Berlin, 1958.

[12] *Code of Practice for Control of Undesirable Static Electricity*, BS 5958, 1991; Part 1: *General Considerations;* Part 2: Recommendations for Particular Industrial Situations.

[13] *Protective Clothing - Electrostatic Properties. Part 1: Surface Resistivity (Test Methods and Requirements)*, EN 1149-1, 2006.

[14] *Electrostatics Discharge Control Handbook*, DOD – HNBK – 263, Dept. of Defense, 1980.

[15] *Recommended Practice on Static Electricity*, NFPA 77, 2019.

[16] *IET Wiring Regulations*, 18th Edition, BS 7671:2018.

[17] H.W. Ott, *Noise Reduction Techniques in Electronic Systems*, Wiley, New York, 1988.

[18] J. Cadick, M.C. Schellpfeffer and D. Neitzel, *Electrical Safety Handbook*, McGraw-Hill, New York, 2000.

[19] *Recommended Practice for Powering and Grounding Sensitive Electronic Equipment (Emerald Book)*, IEEE Std. 1100, 2005.

[20] *Standard for the Installation of Lightning Protection Systems*, NFPA780, 2017.

[21] *Guide for Measuring Earth Resistivity, Ground Impedance and Earth Surface Potentials of a Grounding System,* IEEE 81-2012.

[22] *Low-Voltage Electrical Installations - Part 5: Selection and Erection of Electrical Equipment - Earthing Arrangements and Protective Conductors*, IEC 60364-5-54,2011.

[23] *Protection against Lightning,* British Standards BS EN 62305: 2010.

[24] *Protection of Structures against Lightning*, Parts 1–4, IEC 62305, 2010.

[25] *Recommendations for the Protection of Structures against Lightning*, W.J. Furse & Co. Ltd., Nottingham, UK.

[26] *Lightning Protection System Components*, Parts 1–7, IEC 62561, 2017/2018.

[27] T.M. Kovacic, *Electrical Safety*, American Society of Safety Engineers, Des Plaines, Illinois, 2001.

[28] R.P. O'Riley, *Electrical Grounding*, Delmar Publishers, New York, 1990.

[29] J.D. Cobine, *Gaseous Conductors*, Dover Publications, Inc, New York, 1958.

[30] R. Rudenberg, *Transient Performance of Electric Power Systems*, The M.I.T. Press, 1967.

[31] H.W. Denny, *Grounding for the Control of EMI*, Don White Consultants Inc., U.S.A., 1983.

[32] J.A. Güemes and F.E. Hernando, Method for calculating the ground resistance of grounding grids using FEM, *IEEE Trans. Power Delivery*, Vol. 19, No.2, pp. 595–600, 2004.
[33] T. Takashi and T. Kawase, Calculation of earth resistance for a deep driven rod in a multi-layer earth structure, *IEEE Trans. Power Delivery*, Vol. 6, No. 2, pp. 608–614, 1991.
[34] F.P. Dawalibi, J. Ma and R.D. Southey, Behaviour of grounding systems in multilayer soils: A parameter analysis, *IEEE Trans. Power Delivery*, Vol. 9, No. 1, pp. 334–342, 1994.
[35] C.L. Yaws, *Matheson Gas Data Book*, 7th Edition, Mcgraw-Hill, 2001.
[36] L.A. Lovachev, Flammability limits – a review, *Combustion Science and Technology*, Vol. 20, pp. 209–224, 1979-Issue 5-6.
[37] V. Babrauskak, *Ignition Handbook*, Fire Science Publishers, Issaquah, WA, 2003.

Index

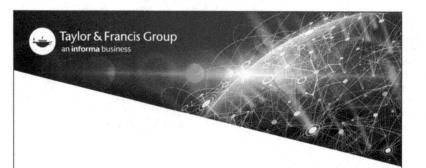

Printed in the United States
by Baker & Taylor Publisher Services

Printed in the United States
by Baker & Taylor Publisher Services